Being
Timeless

Mauro Milita

Linguistic revision and adaptation by Linda Bordoni

*.....to you, who light up my path
even from the shadow.*

CONTENTS

THE DAYS BEFORE AND THE PREMONITORY SIGN   Pg 1

FIRST DAY: THE ENCOUNTER   Pg 10

SECOND DAY: THE ILLUSION OF COMMUNICATION   Pg 20

THIRD DAY: TIME AND SENSORIAL LIMITATIONS   Pg 41

FOURTH DAY: QUANTUM MECHANICS   Pg 78

FIFTH DAY: KNOWLEDGE AND SELF-KNOWLEDGE   Pg 94

SIXTH DAY: THE EXPANSION OF CONSCIOUSNESS   Pg 112

SEVENTH DAY: THE MISSION   Pg 124

BIBLIOGRAPHY   Pg 134

THE DAYS BEFORE AND THE PREMONITORY SIGN

Jan shut his locker with an automatic gesture, an ordinary working day had just ended, but for him it certainly was not like any other day. He stepped out of the restaurant where he worked as a waiter every other evening. He had taken the job, just like so many of his peers, to pay for his studies. Out in the street he found himself bracing against the icy wind of a winter night. As he walked home, a thousand thoughts whirled through his mind. His mother, his only living parent until now, had died just three days before. What would he do now? How much time did he have before he would have to leave the house that, due to a bare-ownership contract, now belonged to someone else? Sure enough, with the little pay he received from his temporary work he could not even afford to rent small flat so, he concluded, he would have to leave his studies and look for a decent job, even if he had almost completed his PhD in Physics.

"Sir, hey, sir" - he heard a voice calling out to him from

behind his back. Jan turned. It was late at night and the only other person in the street was heading towards him at a fast pace. Jan had grown up in a rather rough area and he knew that if he was about to be mugged by a gang, it would serve no purpose to run away. He would be chased and beaten by other thugs stationed further down the road. Staying put, even if with a good dose of apprehension, would at least give him the chance to show he had nothing but a few dollars in his pocket and that he was carrying nothing of value. And if that's what they wanted, he thought, they might as well take it and he would avoid the beating. So he stood his ground and as the young man got closer to him, he noticed that he was coloured, rather slim and walked with a limp. He carried with him a large briefcase that made his gait even clumsier.

"Are you Jan Korpinsky?" asked the boy. "Yes", answered Jan, feeling relieved as the probabilities of a bad encounter seemed to fade. "A man gave me 20 dollars to give you this briefcase, which, he told me, belongs to you".

"No, certainly it isn't mine, there must be a mistake" replied Jan, while in his mind the likelihood of the encounter turning ugly suddenly seemed more probable.

"Look, I don't know what to say", cut short the man, "all I know is that I have to give this to Jan Korpinsky and, if that person is you, I've done my job."

Jan had no time to answer. The boy left the briefcase on the ground, turned and walked back towards where he had come from. The event had shaken him and all his previous thoughts had left his mind. He bent down to look at the briefcase. It was real leather and very well made. It had one of those combination locks with two sets of numbers. It was quite obviously a luxury item, Jan thought, and it had to be very expensive. He clutched it and headed towards the lamp light to get a better look. Just above the handle, there was a brass plate upon which the letters "JK" were engraved. Jan gasped. The initials perfectly matched his own short signature, with the K flowing from the stem of the J in the single stroke of a pen. It must really be meant for him. No mistaken identity it appeared, but who could have sent it? And what was inside? Jan made a couple of attempts to open it by punching in some numerical combinations, but to no avail. It was not one of those standard combinations generally used by the producer for brand new briefcases, such as a series of zeroes or a series of incremental numbers starting with 1.

As he walked home, he couldn't help but think about what had just happened. It didn't seem like a joke. Once home, he placed the briefcase on the floor and paused, sweeping his gaze across the walls of the living room. They looked dull, emptied of all the life they had so far retained, even after his mother had been hospitalized and had not been living in the house for over a month. It felt almost as though the house knew she was no longer here any more and everything that had taken place between those walls over the years had gone forever; it was as if the house were about to be intruded upon by a new owner, who would mercilessly discard all the memories and the legacies of a lifetime, unaware of how much they meant to someone else.

Jan opened the fridge, grabbed a bottle of beer and took out a pack of pre-cooked chicken, which he proceeded to heat in the microwave after having shut the fridge door. He kept thinking of the briefcase and of what could be inside. Why would anyone send something to him without also providing him with the combination to unlock the briefcase, or at least give him some clue to open it? He didn't like the thought of forcing it open but maybe there was no other choice. As he ate his dinner, an idea flashed into his mind: if this mysterious sender knew him so well, perhaps he also knew the number sequence that he most commonly used for passwords, locker combinations and, in fact, for a briefcase.

Aware that this was hardly possible, he decided to give it a last try before resorting to the safe-cracker option. He tried the combination "120" with the first set of three rolls and then "162" with the other set. He flicked the latches up and, incredibly, the briefcase clicked open.

Jan did a double take, both because his code had worked opening the briefcase, and also because of what he found inside: it was stuffed with large denomination banknotes, although they were not grouped into the usual neat bundles but scattered in careless disarray. In the midst of all that money a crumpled advertisement leaflet caught his eye. There were some hand-written words on it that read: "What you believe to be no longer there has not ended but is still there and always will be".

Bewildered, he paused to ponder on the meaning of the sentence, but his mind was empty. Who wrote those words? What did they mean? Why was all that money there? A lot of questions whirled in his mind. He decided to count the money, and after grouping the banknotes according to their value, he found it amounted to some fifty thousand dollars. A huge amount that would allow him to look to the immediate future with peace of mind, at least as far as the financial aspect was concerned. But could he keep all that money? Could he really consider it to be his own? On the other hand, he thought, who could he return it to if he didn't have the slightest clue as to who his mysterious benefactor may be?

He completely forgot about his meal and decided to lie down on the sofa. Immersed in his thoughts he stared at the ceiling until he fell asleep.

His neighbour's damn dog started barking, just as it did every morning at 7 o'clock sharp, when about to be taken out for a walk. Jan woke up and found himself on the couch, still fully dressed, and aware that he had spent a restless night although he could not remember any dream in particular.

He got undressed and stepped into the shower with nothing in his mind but the briefcase. As always, the morning hours had brought with them a different view of life. A feeling of joy and above all a great curiosity had replaced the feelings of distress and dismay he had felt the night before. After all, he thought, except for the cryptic phrase, this was a real stroke of luck. He began to think about what to do next, this new situation would allow him to get organized and to complete his education without having to worry about the immediate need of finding work. But the more he indulged in these thoughts, the more he became aware that things couldn't just come to a sudden end. He couldn't just accept what had happened as a mere stroke of luck. There were too many strange circumstances that intrigued him, driving him to want to find out more. Moreover, due to his rather obstinate disposition, he most certainly would be tormented by them for a long time should he be unable to discover whatever it was behind that whole strange affair.

He decided to make use of the only piece of information he had, well aware that it would most likely prove inconclusive. He would go to that fast-food joint in search of a clue. He got into his car and drove all day long.

The road sped by outside the car, and as the sun began to set, Jan turned off the air conditioning and rolled down the window. He was immediately overwhelmed by the distinct smells of the countryside. He drank in the crispy air with a couple of deep breaths and was pervaded by a feeling of deep relaxation followed by a sharp pain in his tired shoulder muscles where he now could feel the tension of the day. He slowed the car down and pulled to the roadside; outside, the sun marked the end of the day as it slipped behind a barren and stony hill.

With the decreasing daylight, the tremulous lights of a small nearby village appeared in the distance. Jan decided to get some gas at the petrol station there and possibly eat something. He switched on the engine and headed toward the small town. After having left the car at the petrol station, he immediately caught sight of the very same fast-food joint advertised in the leaflet he had found in the briefcase. He walked up to the entrance and pushed open the glass door. A dozen or so people seated at the tables were chatting about the weather and the crops before a drink; a couple of youngsters were trying to coax some music out of an old piece of metal, which had once probably been a jukebox. The face of the young woman serving at the counter was already marked by deep lines; she served him a sandwich and a beer while, each group looked up in turn from its table to peer at the stranger in the hope of finding some element that would do for a topic of conversation regarding the newcomer. Jan did not feel the slightest sense of embarrassment. He picked out one of the small groups, grabbed his tray and walked towards the table, asking permission to join. Perhaps, bored with the long silence and with staring into the bottoms of their glasses, the two occupants of the table readily agreed, expressing their pleasure in welcoming the stranger and getting into a conversation, albeit of a generic nature. Jan did not say too much about himself, but his appearance, which was typical of someone coming from the city, gave much away. So he limited himself to mentioning that he lived in the city and that he was heading south, and leaving his two table-mates to persevere in their lazy chit-chat, he began to eat.

The main topic of local conversation was the burning down of a house, an incident that had occurred a few days earlier, and the fact that the two inhabitants of the house, an elderly couple who hardly ever went out, had miraculously escaped. One of the interlocutors engaged Jan in the conversation explaining how the two old people had been promptly alerted of the flames and helped to escape by an old man called Henry, a strange sort of character who had made his home in a hut on the top of a nearby hill where he lived with his goats. Only rarely did the old man venture down to the village to sell the cheese he produced and to buy something with the money he earned from it. Encouraged by the stranger's interest, they both described the old man as an extremely bad-tempered and reclusive individual, who seemed to be able to anticipate calamities. No one knew what his real name was and in the village he was known as Henry, in memory of an ancient local legend that told of another strange character who, many years before, had settled into that same hut, and who, after some time, had disappeared without a trace.

The first times Henry had happened to be on the site of an accident before it happened, there had been rumours that he was in some way connected to the cause of the disaster. However, something then happened that put a stop to that kind of talk: an old bridge made of timber and ropes, which allowed the crossing of a ditch just outside the village, collapsed while a boy was walking across it. The old man, who was present, had grabbed and held on to the lacerated rope keeping the bridge in place until rescuers arrived on the scene thanks to the boy's cries for help. With bleeding hands, and as usual without awaiting a thank you, as soon as the young man had been brought to safety Henry had made his way back up the hill. Under the circumstances, even although the villagers had by now fully accepted and even respected him, there was always someone who would engage in some superstitious practice or another to ward off possible disaster every time Henry was seen around.

Jan was intrigued by the figure of the old man and began to ask for more information, including indications on how to reach the hill. His companions however hastily advised him against such a visit, pointing out that not only the old man did not appreciate company, much less would he agree to welcome an inquisitive stranger. But Jan took no notice of their words, took his leave, paid the bill and left.

By then it was night, the air was calm and he walked towards the car leaving the faint hum rising from the joint behind him. He got into the car and after having slightly reclined the driving seat, he sat quietly gazing at the clear sky until he fell asleep.

FIRST DAY

THE ENCOUNTER

Jan's awakening was not of the best kind; the noise of a great big tractor passing very close to the car brutally tore him from the first relaxed sleep he had enjoyed in months. The village was already in full morning swing and its inhabitants seemed busy and active although they lacked that spasmodic frenzy so typical of their city counterparts at that same time of day. Jan got out of the car and, stretching his cramped limbs, walked the few paces that separated him from a fountain where he proceeded to refresh himself as best as he could. He immediately headed towards the fast-food/café savoring the aroma of coffee. The woman at the counter looked different, perhaps prettier: her auburn hair was neatly gathered and both her smile and freshly laundered dress suggested she was ready to welcome a new day of hard work. As Jan sipped his coffee, he found himself wondering why he felt so attracted by the figure of the old man. Of course, the story he had been told was unusual enough to justify a certain curiosity on his part, but nothing more, given the fact that he had arrived in that place with a specific purpose. In any case, since he did not have further elements to guide him in his search, he decided to indulge in his curiosity and started to walk towards the hill. The more he imagined his meeting with the strange character whilst climbing the hill, the less he succeeded in finding something convincing to say to justify his visit, any excuse that would appear plausible.

Once at the top of the hill, he walked the few remaining meters before reaching the hut where Henry lived. Two sapling trees, lazily swaying in the breeze, seemed to be contending a string of rags that had been hung out to dry. A couple of steps before the entrance, a chain dangled from a small stack of branches, dipping into a circle of stones laid on the ground. A tree trunk, clearly used as a seat, brought to mind images of long evenings spent in reverie before the fire. He called out and knocked on the door, but no one responded from inside the hut. Jan turned back to the fireplace and sat on the trunk, half-closing his eyes in the breeze. The memory of an afternoon spent in silence with Susan on the bank of a river a few years earlier, came to his mind. Totally lost in his thoughts and in the sound of the wind, Jan sat there for quite a while reliving in his mind the same feelings of that day, until the sound of barking dogs watching over a small number of goats and a man with a tired gait jolted him out of his reverie. Jan stood as if to walk towards the old man who raised an open hand in a gesture which was both a greeting and an invitation to remain seated. Jan slowly sat down again on the trunk while the old man, who had almost reached the hut, called out:

"You are here at last, I was beginning to worry you would

not come."

"Hello! Actually, you see… there must be a mistake… I am not the person..." Jan stammered surprised by those words.

"Perhaps", the old man interrupted, "but it does not matter. Come along, we must celebrate".

Speechless, Jan followed the strange character into the hut.

The interior was covered with huge images of the hut itself immersed in the surrounding wood and portrayed from different perspectives; the images were so perfect they looked real. All around were signs of daily life, like old blackened casseroles, unwashed plates and glasses, outworn clothes, whereas on top of what must have once been a piece of furniture stood a row of excellent framed photographs, all depicting smiling people. The images looked almost real and they were free from that layer of dust and dirt which permeated everything else. Some frames were without photos, while in the lower part of the shelf stood a dictionary; this too was free of dust and presented a glossy cover.

"This is my little kingdom", announced Henry with the tone of someone who has finally reached a resting place, "it's very relaxing, isn't it?"

"It's very original" Jan commented, asking himself whether the old man was really so naive as to not imagine what impression that clutter could have on a stranger.

"What do you find that is so original about it?" chimed in

Henry, showing he had not grasped the irony in Jan's remark.

"Well... for example... I am used to seeing photographs that represent distant landscapes for one to contemplate, I would never have expected to find in... a country house... images of woods and countryside."

"I see, and what else?"

"I am stricken by the fact there is no bed or any other kind of sleeping arrangement."

"What you say is logical, but you will see..." the old man said with a cunning air.

"To tell the truth...", interrupted Jan, "I still need to understand why I have come here..."

"Tell me, I am very interested: how did you get here?"

"Well, at the village I heard talk of a person who lived alone on a hill, and I suddenly felt impelled to meet him...just like that... out of curiosity."

"Where do you come from?"

"Can't you see?" said Jan ironically, "I am an urban animal..."

"No, no, you are right" answered the old man, "you have to excuse me, but the language we speak is of no help, even when it comes to expressing simple concepts. What I meant to ask is from what kind of experiences do you come from? In a nutshell: what is it that generates all that emotional turmoil which is so visible in your eyes?"

"Well...", answered Jan in astonishment, "it must be the stress of city life which is having an effect on me, I certainly do need some time to rest and..."

"You can stay here for a while", Henry interrupted, "if you like".

"The place is truly relaxing, but there isn't even a bed to sleep... as a matter of fact, forgive my curiosity, where do you sleep?" asked Jan, spanning the room from right to left.

"Over there" replied Henry, pointing towards an empty corner, "on a rug".

"You mean that I too would have to sleep on the ground?"

"No, not on the ground", replied the old man candidly, "I have a rug for you too."

"And excuse me: what kind of shepherd are you anyway? You don't look like one at all...in fact, tell me what does a shepherd do with a dictionary?"

"That is my library" answered the old man.

"You mean that you read the dictionary as if it were a book?" asked Jan.

"You see, from a certain point of view, one can say that I am able to read all the books which exist or which will exist, in fact a book is nothing but a combination of the words existing within a dictionary..."

"Very smart...", laughed Jan, both entertained and surprised, "you mean to say that, because a book is a collection of connected words, and a dictionary contains all words, it also contains all possible books?"

"Exactly!"

"You really are a crazy old man!" laughed Jan.

"Stay here for a while, you will see this is a wonderful place, such a beautiful place that... should you wish to decorate the walls with pleasant images, you would not but choose an image of this very place. I mean, if this place contained anything different from itself, it would not be as beautiful" the old man said with a half smile.

Bewildered, Jan leaned on a wall, stared at the old man with widened eyes, and muttered slowly "... I will stay".

Throughout the rest of the day, they walked together in the surrounding woods. As time went by, Jan was pervaded by a growing feeling of inner peace while the nature around them, seen through Henry's eyes and words, seemed to acquire a new meaning, almost as if it had never had that meaning before or he had never perceived it.

There was something fascinating about the old man, something that did not really fit in with his appearance: he looked like a shepherd, but his language was too polished. It revealed remarkable culture and depth, even although up until that moment he had expressed himself using only basic images.

As they walked, they reached what Henry called the "source", an iron pipe fixed into a cleft in the rocks from where extremely cold water was flowing. They drank eagerly quenching their thirst and then sat on two large stones which seemed to have been placed there for that purpose. As the day made way for the evening hour, the vivid colours of the countryside slowly faded to grey.

"We should go back home shortly" said Henry.

"Home… what a different meaning this word has for the two of us…" Jan murmured.

"That depends on how each of us has known it" replied Henry.

"What do you mean? Why do you always use such strange expressions? You come across more as a philosopher than as a peasant… actually… as a deranged philosopher!" the young man burst out, tired of the other's elusive talk.

"You see, knowledge is a complex thing… if you forgive my game of words, I would tell you that you do not know it. The act of knowing is so ultimately beautiful that it cannot be further embellished if not by directing it towards itself, which means… you will never be able to know anything if first you do not know knowledge", warned the old man.

"Just like the photographs of the house in the forest inside… the house in the forest?" asked Jan with wide-open eyes.

"Exactly!" answered the old man, almost crying out with satisfaction. "…It is actually you… but enough now…" he cut short, "let's go back".

They stood and headed towards the hut without speaking. As Jan delighted himself in listening to the sound of his steps treading the grass, he suddenly realized that the old man's contorted reflections had made him forget his problems for quite a long period of time.

Once at the hut, it occurred to Jan that he had to go to his car, which was parked at the bottom of the hill to fetch his bag and the clean clothes he had brought with him. Most certainly, the old man did not possess a torch so he should hurry before night fell.

He walked with a fast pace down the hill and returned to the hut when it was almost dark. The building's shape no longer conveyed that sense of instability which was so evident in the daylight. Looking at it now, Jan had the feeling of a comfortable shelter in the middle of the dark countryside. He felt hungry and realized that he had not eaten anything after his morning coffee, he hadn't event thought of food all day but his appetite suddenly made itself felt.

Once inside the hut he came across the old man hustling around the old furniture which was falling to bits, but nevertheless it was clean. He took out some canned beans and a small piece of cheese.

As he watched the old man's gestures which betrayed the tedium of an every-day routine, Jan could not help reflecting on the fact that his presence did not appear to unsettle the old man at all, and that food, for him, was most probably of secondary importance. An oil lamp shed light on the wooden table and on a modern-looking chair that was in such bad condition it seemed to have been found in a dump.

The old man dragged a tree stump which he used as a stool towards the table, sat on it and pointed his guest to the chair. They consumed their frugal meal almost in silence and when they had finished Jan pushed his chair away from the table and watched the old man, who was staring into space.

"How long have you been living here?" asked Jan.

"Who can say? I feel as though I have always been here" answered Henry.

"...and where were you before? ... What did you do?"

"What does 'before' mean..." asked the old man in an annoyed tone "before I was... no... I am... well, I have been a professor".

"There you are!" cried out Jan in a flash "I knew you were not a real shepherd! But how come..." and interrupting his sentence he went on: "...perhaps that is something you do not want to talk about?"

"I do not mind speaking about it, I just consider it irrelevant to know who or what I was before, because 'here' and 'now' are something else" answered the old man.

"But that is not true!" replied the young man, "To me your words are useful in order to understand..."

"What is it you wish to understand? And words? It is not easy to learn from words. They respond to an incomplete logic" asserted the old man, almost in a huff.

"Which is?" asked Jan.

"It is too late now to explain how complex it is, but if you are interested in the journey to understanding, you can stay here a while" said the old man.

"Why are you doing this?" asked Jan, intrigued by the offer.

"Because each one of us should pay tribute to intelligence when he encounters it, and this is my way to honor it".

"I see" answered the young man, "but I do not know whether I will be able to survive your cooking."

They both laughed and then the old man took a well-rolled up mat out of the cabinet, handed it to his guest and pointed to the place where he could lie down and sleep and also to a small closet that could hardly be called a bathroom, but which was ingeniously equipped to provide for basic needs and to serve as a sort of shower.

Lying on the mat, with the lights off, Jan looked within himself. The anguish, which had overcome him until just a few hours before, had magically vanished and he was both intrigued and fascinated by that uncommon character. He did not feel the slightest wish to go back to his world that now seemed rather useless. Tired, he fell asleep with the shadow of a smile on his face.

SECOND DAY

THE ILLUSION OF COMMUNICATION

Jan's awakening was very bad as he was not used to resting in those conditions and his body felt full of aches and pains. Daylight flooded the hut's interior, while a regular banging noise could be heard coming from outside. He felt so stiff he would easily have been able to draw up a detailed list of all his muscles and bones. Grimacing in pain he got up, attempted to stretch his aching joints and dragged himself to the door. Henry was busy chopping wood with an axe, while a pan full of milk was placed over the fire on a tripod.

"Good morning," Jan croaked.

"Good morning to you", replied Henry straightening his back and turning towards his guest. "Did you manage to get some rest?" asked the old man with a smile that said he already knew the answer.

"I slept like I have not slept in a long time" answered Jan

trying to stretch his muscles, "even though I feel a bit sore".

"Your brain finally got a moment of rest and did not control your body during the night, leaving it to itself" answered the old man, "that is why you are sore".

Jan chose not to reply and explain all the advantages to be gained from a good mattress; he limited himself to wry smile and turned towards the bathroom.

Henry sharpened his axe on the stump and headed to the hut. Meanwhile Jan tidied his things.

"It's nice, finally, to have breakfast in company. I have prepared some fresh milk and bread", said the old man.

The two sat at either side of a rickety board upon which Henry had set out the few things he possessed perfectly rendering the idea of a well laid table. Jan realized he was hungry and soon forgot the initial disgust he had felt because of the less than hygienic conditions of the food and the strong taste of the milk.

He ate with gusto and then, looking at the old man, asked: "I have thought a lot about what you said yesterday regarding the inadequacy of words to communicate. Even though it is not easy to convey to others exactly what we intend to say, we can help ourselves with gestures, facial expressions, images, making our message more univocal."

"Come, it's a beautiful day! Let's go for a walk together and we will talk about communication," Henry proposed.

The two slowly walked down a lane leading from the hut into the forest.

"See," the old man said in a low voice "I am going to ask you a question: is it correct to say that a statement can be either true or false?"

"Of course! What else could it be? If it is true it is not false and vice-versa."

"Good, then how would you classify the sentence: this sentence is false?"

"Well," Jan stammered after a moment of silence, "...if the sentence is true...then it means it is false...whereas if it is...But this is a game of words!" Jan protested, irritated by the impasse he had got himself into.

"It could be," winked the old man, "but as you see we have pushed the set of words to the edge of a cliff: a black hole in which logic is lost and what you say seems to have no meaning. The paths of knowledge are other ones, they do not pass through words if not for brief and fleeting moments in order to provide support to thought, but you need to be very careful as to where they rest.

"What do you mean?" the young man asked in admiration and wonder.

"It means that the concepts expressed by words can build true knowledge only if their foundations are sound, that is to say, if one is far enough from the black holes of logic. In fact meaning cannot be considered as something intrinsic to the message itself, but the product of the interaction between the mind and the message."

"I see. So in order to have good communication it is not enough to work on a good message but we need to keep into account that he who receives the message can give it a very different interpretation to what we expect", said Jan.

"Certainly. The sentence I quoted, "this sentence is false" represents a paradox, that is to say, a black hole of logic in which our ability to use reason comes to a halt because we do not have, at least apparently, the tools to tackle it. The main obstacle is quite hidden. It lies in the fact that that sentence speaks of itself, it is self-referenced… it is closed upon itself like a ring and does not leave us any space in which to explore it with reason because it does not have a starting point nor a conclusion" said the old man.

"I understand. The sentence is closed as if in a circle and, if we walk in a circle, we find ourselves at the point of departure. " Jan agreed.

"Exactly. When we try to get into the sentence using reason, we find ourselves at the point of departure, just as in the famous lithograph by M.C. Escher known as "Ascending and Descending". In it, two rows of people respectively climb up or down a staircase that is closed in itself with ends that meet. Amazingly, whether they walk up or down, they always find themselves at the point of departure.

"The analogy between the sentence in question and the work by Escher is very appropriate" said Jan.

M.C. Escher: Detail from the lithograph "Ascending and Descending"

"Exactly. One of the first known examples of such problems in logic is known as Epimenides Paradox. Epimenides was a Cretan philosopher who pronounced the following sentence: All Cretans are liars... consequently so was his sentence which, if examined by the same standards as this sentence is false, leads to an endless game of mirrors. But a ring can also be formed by joining two semi-circles...that is to say...there can be some assumptions that if taken singularly are harmless but, once put together, become consolidated into a logical circle.

For example: "The following sentence is false – the previous sentence is true." We need to be careful in our arguments until we get to know the rules that will allow us to avoid falling into the black holes of reasoning, We also need to avoid these pitfalls as we put sentences together, sentences which, if taken independently, appear harmless" said the old man.

"Right!" agreed Jan.

"Now let's take a step forward and speak of language. That is singular isn't it? We intend to describe language by using language itself. This is also a logical ring, or, as it is commonly known, a strange loop. It is one of those cases where container and content are confused. This situation is well described in the famous question: can a vase contain itself? This condition is rather misleading because, even if well hidden, it contains the concept of infinity, which generates a loop in the mind of he who faces the problem.

We often face strange loops in everyday life without even realizing it as we automatically push away the logical problem contained therein. Here too, our understanding can be aided by a lithograph by M.C. Escher called **Print Gallery**, Henry explained.

When you will have the chance to observe it, you will most certainly stop for a moment to figure out what it is you are beholding. In fact, the observer's mind needs to ask itself whether the work represents and includes itself, or whether it is a work that represents a gallery of prints that includes itself or any of the other possible interpretations.

M.C. Escher: Print Gallery

In fact, in the presence of a logical ring, we can never be sure we have taken the right path. Therefore, getting back to our intention to speak about language, we need to accept the fact that it will be an incomplete and patchy analysis", concluded Henry.

"So" Jan continued, "it is important that the mind seeking to grasp a meaning does not get lost in the idiosyncrasies nestled in the very structure of the message".

"Certainly, but that is not all. It isn't sufficient to consider language only, the issue is more general. It's often very difficult to understand if something has a message at all, let alone understand its meaning".

"But because the meaning is contained in the object, it should be sufficient to analyze it…" objected Jan.

"That is not so, my young friend. Think, for example about deciphering a text from an ancient and unknown language. Simply by observing the series of signs, we can derive that it is a result of the intellect and therefore it must contain a meaning. However, even if we manage to decipher its meaning, how can we be sure that it exactly reflects everything and only what the author intended to communicate?

"How can we establish whether our understanding, that is to say, the interaction between the message and our mind, has not impaired its content in any way? Comprehension is always the fruit of the combination of mind and message, and it cannot be attributed to the latter only", said the old man.

"But the problem is that the previous sentence described itself", objected Jan.

"True, the ability to go outside the context is the most

difficult part but is also the solution to the problem", Henry announced, pleased with his interlocutor's ability to not be frightened by logical paradoxes. He continued "This depends on the ability to aggregate reality into categories, an activity that is directly proportional to one's level of intelligence. Bees, for instance, seem to possess an innate idea of the flower category, however the same does not apply to everything else that surrounds them. Going up the intelligence ladder, dogs and cats seem to be able to formulate new elementary categories, but only man appears to be capable of identifying himself as being part of a category.

In fact, this is possible only if you look at yourself from outside and notice, for example, that you possess a head even if you cannot see it, simply by observing other representatives of the human category", the old man concluded.

Jan listened in total rapture to those quiet words from which enormous self-confidence transpired.

"The capability to step outside the context," said the old man, "is, you must remember, the key to understanding. In fact, if to understand literally means to embrace, to enclose, how can you do so if not from the outside?" said Henry, stressing his words with a raised finger, just like an experienced professor.

The simplicity of the arguments was so disarming that Jan wondered how he had never thought about such things in spite of touching on similar topics several times.

"To communicate exactly is not just difficult… it is practically impossible, and this is not just due to the inappropriateness of words, communication itself is a problem per se", continued the old man.

"What do you mean?" asked Jan with curiosity.

The professor got up and reached out to a shelf where he took a pencil and some sheets of paper that had turned yellow with time. He sat down again and sketched a half open door in the middle of the paper. "What is this?" he asked, looking at the young man with a soft but challenging tone.

"A door" answered Jan.

"Good, if that is a door… then open it!" demanded Henry.

Jan was speechless; he did not understand whether the old man was joking.

"Obviously you cannot open it… and why can't you?" carried on the old man.

"Because it is not a real door …" Jan began.

"Of course…" Henry interrupted, and went on "it is not a door but a drawing, that is to say it is only the representation of an object or of a concept that I had in mind and wanted to communicate…could it be for example that maybe I did not wish to communicate the message door but rather the message exit...?" asked the old man.

Jan, who kept staring at the sketch, had to agree "...yes actually that half open door could also mean exit...".

"See..." carried on Henry, "the mechanism of communication is complex, and we do not realize it as we speak, but it makes our communication very difficult. Now I am going to show you a concise analysis of the process...".

Jan was surprised; not only because of what he was hearing, but also because he no longer beheld a bizarre, scruffy old man, and the conversation was taking the form of a lecture. In spite of his gentle and friendly tone, the old man's words were assuming an authority and strength that left him in wonder.

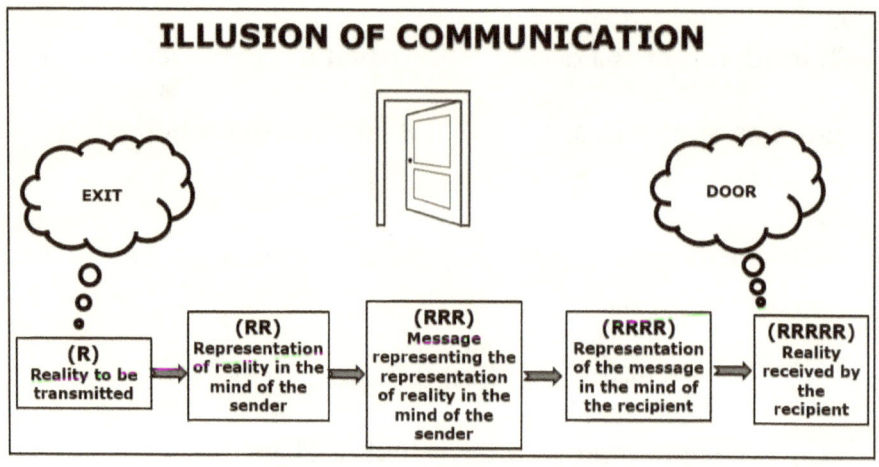

"We have said that, in my perception of reality, I take the exit concept and, in order to communicate it to you, I build its representation in my mind, and it is logical for me to imagine a door. Then I transfer and enclose my mind's representation into a message, the drawing. But my message is not the reality I want to communicate, it is just my mind's representation of the reality.

When you receive and read the message, you build in your mind its representation and through this, you create your own representation of reality. Therefore, as you can see, after all these subjective representations, how can you be certain that the original reality I wanted to pass on to you actually reached your mind unchanged?" the professor asked in a provocative tone.

Jan listened open-mouthed while his eyes followed the sketch those rough hands were drawing on the sheet.

"So then… this discussion of ours is quite useless too… how can you be sure that I will grasp the true meaning of what you want to communicate with your words?" asked Jan.

"Well… it is not completely useless…" answered the old man, "we will manage to communicate well enough if we keep in mind the barrier posed by words and if we complement these with questions and answers, in order to adjust their relevance to the content. This is what I wanted to tell you right from the beginning."

"Fair enough. Let's try", said Jan.

The old man carried on: "You see, historically we have often relied only on words in our process of comprehension. This is one of the reasons why in the past, even while acknowledging the intellectual greatness of thinkers, philosophers, writers or scientists, we have limited ourselves to examining their texts without allowing them to blossom into fragments of higher consciousness and knowledge.

Thus, their enlightenment has most probably been lost to us to be recaptured only when the objectivity of science has led us back to those same concepts.

Some of the important topics we are going to speak of are related to the concept of time and the number of dimensions that limit our perceptions. Many of the things we will talk about may sound new to you, as they are milestones that science has only recently reached.

For example, in 1879 Dostoevskij wrote something enlightening about the difficulty to perceive multiple dimensions, but what he was saying was perhaps not sufficiently investigated, nor did it penetrate our consciousness.

"...if God exists and if he indeed created the earth, then, as we know perfectly well, he created it in accordance with Euclidean geometry, and he created human reason with a conception of only three dimensions of space. At the same time there were and are even now geometers and philosophers, even some of the most outstanding among them, who doubt that the whole universe, or, even more broadly, the whole of being, was created purely in accordance with Euclidean geometry; they even dare to dream that two parallel lines, which according to Euclid cannot possibly meet on earth, may perhaps meet somewhere in infinity. I, my dear, have come to the conclusion that if I cannot understand even that, then it is not for me to understand about God. I humbly confess that I do not have any ability to resolve such questions, I have a Euclidean mind, an earthly mind, and therefore it is not for us to resolve things that are not of this world. And I advise you never to think about it, Alyosha my friend, and most especially about whether God exists or not. All such questions are completely unsuitable to a mind created with a concept of only three dimensions".

Dostoevskij: Brat'ja Karamazovy (The Brothers Karamazov)

Many centuries before him, St. Augustine, in his Confessions, had clearly written of the nonexistence of time and its illusory perception deriving from our limitations. But even in this case, such great enlightenment, although much valued and appreciated, was considered mainly only under the religious-mystical-philosophical profile".

But what now is manifest and clear is, that neither are there future nor past things. Nor is it fitly said, «There are three times, past, present and future;» but perchance it might be fitly said, «There are three times; a present of things past, a present of things present, and a present of things future.» For these three do somehow exist in the soul, and otherwise I see them not: present of things past, memory; present of things present, sight; present of things future, expectation. If of these things we are permitted to speak, I see three times, and I grant there are three. It may also be said, «There are three times, past, present and future,» as usage falsely has it. See, I trouble not, nor gainsay, nor reprove; provided always that which is said may be understood, that neither the future, nor that which is past, now is. For there are but few things which we speak properly, many things improperly; but what we may wish to say is understood...

At no time, therefore, had You not made anything, because You made time itself. And no times are co-eternal with You, because You remain for ever; but should these continue, they would not be times. For what is time? Who can easily and briefly explain it? Who even in thought can comprehend it, even to the pronouncing of a word concerning it? But what in speaking do we refer to more familiarly and knowingly than time? And certainly we understand when we speak of it; we understand also when we hear it spoken of by another. What, then, is time? If no one asks me, I know; if I wish to explain to him who asks, I know not. Yet I say with confidence, that I know that if nothing passed away, there would not be past time; and if nothing were coming, there would not be future time; and if nothing were, there would not be present time. Those two times, therefore, past and future, how are they, when even the past now is not; and the future is not as yet? But should the present be always present, and should it not pass into time past, time truly it could not be, but eternity.

St. Augustine: Confessions

"You mean that we have had precious hints before our eyes which we have never sufficiently understood or appreciated?" asked Jan.

"Exactly. That is also due to the fact that communicating through words that are duly connected into a structured language, not only is not at all simple or univocal, can hardly lead to the activation of higher channels of knowledge and consciousness. Therefore, seeing that language that is composed of words and structural rules is somewhat insufficient in conveying the true content of messages, we can ask ourselves whether something more powerful exists in order to reach that scope."

"Do you mean something different from language or are you are talking about a more powerful language?" asked Jan.

"Well, maybe both things. First of all, let us attempt to define what I mean by power of language.

I believe a good definition might be its capacity to transfer content compared to the numbers of components that have been used. In fact, street signs can be considered more powerful than spoken or written language in conveying a message that would require many words.

For instance, the sign that indicates the presence of a slippery road consists in a few symbols, but if we were to describe the message it contains using words, we should say it warns of: the risk of losing control of the  car's movement in relation to the driving speed inasmuch as the wheels' grip on a tarred surface is inferior to normal" the old man said.

As you see, I have had to use many words, even just to adequately define the adjective "slippery". On the other hand, through symbolic language, I have managed to convey the message more immediately, using a much lower number of components and above all by eliminating the ambiguity of words, almost completely.

So one could derive that language is all the more powerful when it succeeds in conveying a message using the smallest number of elements, thus decreasing the number of potential ambiguities" concluded Henry.

"Yes…" said Jan, "I had never thought about street signs as a form of language, but in fact that is what they are."

"Good, now that we have seen how there can be many languages, each with its own power, we need to understand that the messages we wish to convey can be more or less simple and in function of this we should search for the right form of language that allows us to convey its meaning in the best possible way."

"What do you mean by messages that can be more or less simple?" asked Jan.

"In the previous example the message referred to and described something visible, tangible and physically verifiable, like a car losing control. But if I wished to communicate something that is more complex and not directly visible or tangible, such as, for example, a feeling, an emotion, a mood, or something that cannot be sensed by the five senses, what would the most appropriate language be?" asked the old man.

"Well that is something that I can communicate through words, in fact I can say: I feel sad, or I am furious or something similar", objected Jan.

"Of course," answered Henry, "but because these are subjective feelings, referring to what we said before concerning mental representations, you can never be sure if the meaning of the word sad is the same for everybody, if not in very general terms. In fact, if you wish to be more specific you would have to start using many more words in order to convey what it means, for you, to be sad".

"True, I see what you mean. For example, a painting or a drawing can represent that message with more power, just like the street sign does with respect to words. " observed Jan.

"Exactly, but the picture or painting is just one of the possible channels. Think about music and the way it can convey feelings without using words."

"True", observed Jan, "music is certainly a very powerful language, though I would not know what sort of structures it works on within the listener's mind."

"Then you should think that these are nothing but tricks to get around the fact that our brain's representations often lead us astray when we try to understand a message. If you follow me carefully I will reveal to you the hidden trick behind this kind of communication."

"I am all ears. " said Jan.

"Alright, paradoxically, the real problem is the very structure of language. In fact, the brain gets lost in the inevitable ambiguities of interpretation of the structure and loses sight of the original meaning. Sometimes it even reaches a real structural deadlock as in the case of the paradoxes of which we are going to talk about shortly."

"Well then?" asked Jan with curiosity.

"The senses should receive a stimulation that is as destructured as possible, so they can transmit their stimulus to the brain which can then choose from its own representations. For example, a painting or a drawing makes use of the sense of sight to reach a recipient and in the blink of an eye, can convey a great number of feelings or emotions that would be very hard, if not impossible, to convey through words. The same happens when we listen to music plunging ourselves into a whirl of emotions, or when we smell an odour or a perfume that intoxicates us, or even when we react to the touch of a surface that is soft or jagged, wet or dry, flat or curved. Technically, each sense can be used for a specific language, better still when we can use more than one sense at a time. For example, a film can stimulate both sight and hearing simultaneously."

"I understand," said Jan "basically you mean that we can represent something that eludes the senses by using our senses, if we use the appropriate languages."

"Yes, but be careful, because our senses can cheat us."

"What do you mean?" asked Jan.

"Well if we take for example sight, everybody knows it can send us the wrong information, such as the stick that appears broken in half if partially immersed into water, not to mention the many examples of optical illusions. The same applies to all of our senses, therefore we need to know that communication, and more generally the perceptions we receive through our senses, can be highly misleading."

"Then why did the good Lord play this trick on us? Why did He provide us with imperfect senses that confuse us both in communicating and in perceiving reality? That is wicked. " said Jan teasingly.

"On the contrary, this constitutes a precious help, a powerful indicator that informs us that if we really wish to gain a deeper understanding of the reality we belong to, we need to come to terms with our own limitations, avoid being misled by our senses, and learn to communicate and perceive in different and more complete ways."

"Beautiful words," said Jan "but what do you mean in practice? What should we do or invent?"

"In practice… in practice…" repeated Henry "…the eternal obsession. You will never arrive anywhere in this way. You need to abandon phenomenal logic and think that there is a different form of communication and perception. Although you may not be able to see it at the moment, it is enough to think that it can exist. The time will come when you will perceive it inside yourself even if perhaps you will never know in which way it came to you."

They had by now reached the end of the path where thereafter began a much thicker part of the forest. Thus, they turned and walked back towards the hut while continuing to talk about how important it is to give value to those achievements of the mind and the spirit, which do not appear to bear immediate consequences on the material world. After consuming a frugal dinner, they wished each other goodnight and retired for the night.

THIRD DAY

TIME AND SENSORIAL LIMITATIONS

The increasing brightness of daylight filtering into the hut through its rickety ceiling gently woke Jan up. He opened his eyes and slowly began to look around. The feelings he felt regarding the environment had changed compared to the first day. Everything seemed less strange and, albeit in a confused manner, his mind could perceive there was a meaning in the objects that surrounded him and in the way they were arranged, even though he could not say what it was.

"Good morning" greeted the old man as he entered the hut from milking the goats.

"Good morning" said Jan in return as he got up with a smile that took even him by surprise.

After a light breakfast, the old man placed a piece of cheese in front of him and said: "Look at this, can you see it?"

"Of course!"

"I don't think so. You are convinced you can see it, but as a matter of fact, it is only its surface that you are seeing. Moreover, even if you can handle it and understand that its weight does not derive only from the surface, you can still neither see it completely nor touch it completely except in its external part."

"What do you mean?" asked Jan with curiosity.

"You see, because this object is solid, its entirety eludes us into a higher dimension because it is closed in itself, just like, from a logical point of view, the famous statement: **this statement is false**."

"I understand, you are saying that I cannot see or touch the parts inside. But I can always cut it to access them, can't I?"

"This does not change things. For as much as you cut it and divide it into thinner slices, there will always be an internal part of that slice that you will not be able to see or touch. " said the old man.

"And does this depend on dimensions?" asked Jan.

"Certainly. If there existed a being able to perceive four dimensions he would see both its surface and its inner part at the same time."

"So, if I understand correctly, I am condemned to not be able to actually see, completely, anything of the world surrounding me except its surfaces?"

"Exactly. You see, when our being is imprisoned in a physical body (some say cyclically), it is subject to all the limitations of that body. And, in the case of man, these limitations are so heavy that they have held back man's progress for centuries without him even realizing it. Human memory starts from zero (except in some rare and strange cases) and man is subject to the limitations of a physical body, which do not allow him to perceive more than three dimensions. From this limiting position, man has always developed his own idea of the world, and has always had great difficulty in accepting differing evidence that has been revealed and proposed by science time and time again.

For a very long time, for example, it seemed logical to think of the world as flat. The earth could not be perceived as a sphere because of its extremely long radius and man could not conceive the fact that the ground beneath his feet was curved whilst he saw it as being perfectly flat."

"And how could we have guessed, without basing ourselves on our senses that, for instance, it was not the sun to revolve around the earth but the opposite?" asked Jan.

"Science eventually put us face to face with our limitations. We have now understood that our eyes can only perceive a tiny part of the radiation of light, our ears a tiny portion of sound waves and so on, and we have learnt to operate within entities, measures and dimensions, which we can't even imagine. Think of imaginary numbers (commonly used in electronics), of isotropic lines (perpendicular to themselves) and so on.

Our inability to fully comprehend the environment in which we live has even been scientifically expressed by Heisenberg who described it in a principle."

"So you are saying that science has helped us to correct wrong or incomplete ideas that we have established, through the centuries, based on our senses?"

"Yes, and it is still helping us, at times even changing some of our most deeply rooted concepts. For example, **the greatest deception we experience relates to our perception of time**. If we try to think what time actually is, we invariably come up against a logical wall. We think that time exists, in fact we have even invented many methods to measure it, but in reality it is something very strange."

"Strange? To me it seems that there is nothing more natural and real than time" asked Jan with wonder.

"You see, we are used to dividing time into three big parts: the past, the present and the future. Now, if we were to express ourselves based on appearances we should say, in total agreement with St. Augustine, that the past is something that has existed and no longer exists, whereas the future is something that does not yet exist, and therefore in the present, neither of them exist."

"True..." said Jan, who could not deny the logic behind those statements.

"So, if the present is something that separates the past from the future, we are definitely speaking about something that, by its own nature, separates two things that do not exist. Quite an absurdity, isn't it? And then, what *depth* should the present have? Just because you can divide it endlessly, to the most infinitesimal unit you can conceive, can we actually say it exists? When does past actually begin? A billionth of a second ago? Or even less? As you can see it is pure illusion, a trick of nature to compensate for our incompleteness."

"You mean that neither past nor future exist and, from what you are saying, not even the present?"

"They do not exist in the sense we commonly understand these terms. The past is not over and the future has already happened. Try to imagine them as if they were located, respectively, in different places in space, and think of the present as a journey between these places. In fact, during a journey, the town we leave behind certainly does not stop to exist when we cannot see it anymore. In other words, you must extract from these terms the concept of time that alters our comprehension of reality."

"You mean that to conceive events in time is an obstacle to our comprehension?" asked Jan.

"Exactly. For too long we have believed in the existence of time as something specific and unchanging and we have created hundreds of ways to measure it, building all physical laws upon something we believed to be a pillar of the universe. Only when Einstein with his Theory of Relativity proved that its magnitude was variable and only apparent (so much so that it varies according to the variation of speed of every individual observer), even some of the most orthodox scientists began to review the concept and got closer to the positions of philosophers who had sensed its ephemeral nature."

"So, even in this case, we have not really understood the message that philosophers wanted to communicate to us." said Jan, who was now totally tuned-in to the question.

"At least we relegated it to the role of bizarre reasoning until science put before our very eyes the way it actually is. In any matter, even understanding that it is all an illusion, it's man's predicament to 'keep to the game' of nature, almost as if he had to pretend to believe in it".

"So we have used our science for centuries, developing laws and principles without knowing they were all based on illusions. But then how is it possible that these principles work, that laws give the right results and that, based on this imprecise science, we have managed to build our progress and achieve ambitious goals?" asked Jan.

"You see, even if we understand that the measure of time varies with speed, because relativistic variations become relevant only at speeds that are close to that of light, our science continues to use Newtonian physics for common activities, and only in particular cases is it necessary to use laws that take into account relativistic corrections."

"So, even knowing that time is variable, in everyday life we are bound to consider it fixed and unchangeable and on this we base a good part of our life. How funny" commented Jan to himself.

"Exactly. Moreover, even today, we do not have univocal answers, in the world of physics, regarding the exact nature of time. In this regard I do however have a precise idea, which might seem quite subversive, but it provides a coherent explanation to experimental results and phenomena which would otherwise appear in contradiction or simply incomprehensible" said the old man.

"I am curious to hear it." said Jan.

"Well, the detached observation of all that surrounds us and a growing state of consciousness which allows us 'to see things from the outside' tells us, without doubt, that time simply **DOES NOT EXIST**.

"What do you mean?" asked Jan in wonder.

"I am going to use an analogy to help you understand this better. When you wish to photograph an object that is larger than the width of the camera lens' field, and you cannot change your position or that of the object, you are forced to photograph it piece by piece. Isn't that true?"

47

"Yes, I see, you mean that in this way you spread across time the acquisition of an object which you cannot perceive entirely."

"Exactly. But it is better to turn this paradigm around and to state: **we are forced to decompose in time all that exceeds the dimensional perceptions of our senses".**

"It's true. I had never thought of it in this way. " said Jan.

"So, by imagining that everything that surrounds us has more than three dimensions, we can be sure that we directly perceive only that which takes place within the space of the first three dimensions and the rest escapes us, or more precisely, we perceive it in a form that is distorted in time.

"Incredible. Then we lose a lot of the world around us!" exclaimed Jan.

"That's right. Generally speaking, we can assume there is a sort of fundamental law which states:

All the intersections of our world with realities that present more than three dimensions are perceived by us as phenomena which vary in time."

"Then can we assume that everything is in space?"

"Yes, in a way, even though we speak of a space with more than three dimensions. If that law were universal, that is to say valid for all numbers of dimensions, we could derive from it the following statement: **Time does not exist in the way we perceive it but it is only the manifestation of a space of a superior order.**

Or, more exactly: **in the observation of a phenomenon of multidimensional nature, the articulation of that phenomenon in time is only a sort of deformation by which we perceive its extensions into dimensions that are beyond the third one."**

Jan's eyes were wide with wonder. The apparently elementary affirmations of the old man and the persuasive logic of his reasoning were giving him a completely new and different view of the world. He sensed that from those considerations enormous consequences ensued concerning the perception of the surrounding world and the image that man has of himself. He also felt that he was not immediately able to estimate all the implications and that he needed to allow those considerations to sediment in his mind in order to be able to assimilate them slowly and grasp their depth.

"This means that…" murmured Jan, while his eyes, gazing at the sky, seemed to follow an imaginary course.

The old man observed him with a smile and kept pointing out: "Therefore there are no *yesterdays* and *tomorrows* as we commonly understand them, and what appears to have happened yesterday has not ended and can still be found somewhere, just like that which appears to be going to take place. They are located in spaces that we cannot reach. But this does not imply that the future is already preconceived because, as we will see later, there is an infinite number of possible futures, even if all of them are already present. And the different realities in the multiverse are all diverse and yet verified, in the same way as a single electron can simultaneously be present in different places in space, as based on the intuitions that quantum mechanics has provided us."

"You mean there is a place in space-time in which we are still children and another in which the body we occupy is already dead?" asked Jan.

"That is right. We do not have memory of the first and we cannot yet see the second, however they already exist in a specific place."

"Now I understand", said Jan "when our physical body dies and we lose its limitations, should our mind continue to exist, all this would appear to us as obvious and accessible."

"I do not know. However, what is clearly singular is the fact that stories about a phenomenon called NDE (Near Death Experience), that is to say the testimonies of people who have had a near-death experience and who, for whatever reason have come back to life, do present a common denominator. They all say they were able to see their entire life span, with all its important moments, as well as their loved-ones who had passed away. They also say they were able to observe their own bodies, lying there as if dead, from some sort of an external view point. In other words, they were able to perceive events from a place outside time, a place where those moments never ended but are still 'there'. In a nutshell, it is as if they were able to get "outside the context" said Henry.

Jan was now feeling more confident, full of enthusiasm for the series of intriguing images that kept forming in his mind that to him seemed like real enlightening revelations. As he continued to daydream, he said:

"This means that even the Christian paradigm could be read in the light of these conclusions. In fact, Adam, a perfect being in a multidimensional world (i.e. paradise) could have been punished by being subjected to a three-dimensional universe and consequently condemned to live events of a superior kind perceiving them in time. With death, as at the end of time, such limitations disappear and he acquires consciousness that past, present and future are only one dimension in a superior space and that they exist simultaneously. God therefore, but man too, will be able to see them all during the last judgment".

"Well, it makes sense" said Henry, "Now to better visualize these concepts we can use the same example described by Hinton in the book *The Fourth Dimension*."

"Yes, please describe it." said Jan, now fully immersed in the conversation.

"In order to visualize how the number of dimensions limits our perception, considering that we cannot perceive a fourth dimension, let's try and see how a hypothetical bi-dimensional being perceives a three-dimensional object. Hence we will attempt to understand what could be our condition with respect to a fourth dimension."

"Yes, a good trick" said Jan.

"So, our hypothetical flat character lives, obviously, on a flat plane, moving within a three-dimensional space, although, naturally, he is not aware of it.

When his flat world interacts with a three-dimensional object, for example a spring, he will not perceive the existence of the object but only the effects of an intersection with his world, which vary in time.

As the flat plane and the object move vertically with respect to each other, their intersection will describe a circle. Obviously, not even the circle can be perceived by our flat character in the same way as we see it, because, in order to observe it, he would need to be elevated above the flat plane he is bound to, and this is not possible. So he will perceive a spot that moves in time along a closed curved line and he will be able to develop his own totally valid scientific theory, which includes a law according to which this spot moves."

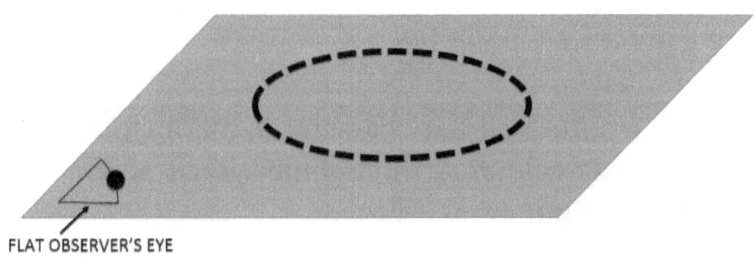

FLAT OBSERVER'S EYE

"So you are saying that if we transpose this small example into a wider context, we can affirm that we are not able to perceive objects that have a higher number of dimensions than ours but only the effects produced by their intersections with the three-dimensional world." said Jan.

"Exactly. So we could say: **an object existing in a space with *n* dimensions cannot be perceived by an observer with *n-1* dimensions if not as phenomena that vary in time.**"

While Jan pondered on those concepts, trying to visualize them in his mind, the old man went on: "Now let's extend this example a little to observe in greater detail what the implications are regarding the time aspect.

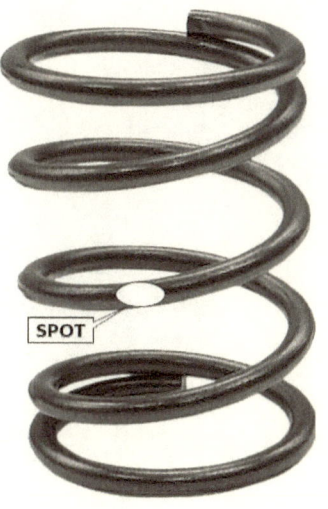

Let's imagine there is a particular spot on the spring, like a variation in its shape, or a simple discontinuity that produces a variation in the perception of the observer.

He will see the effects of this disruption disappear slowly as the flat plane and the spring move in respect to each other.

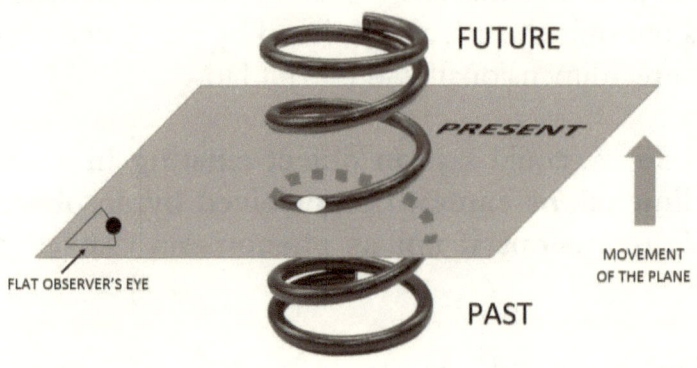

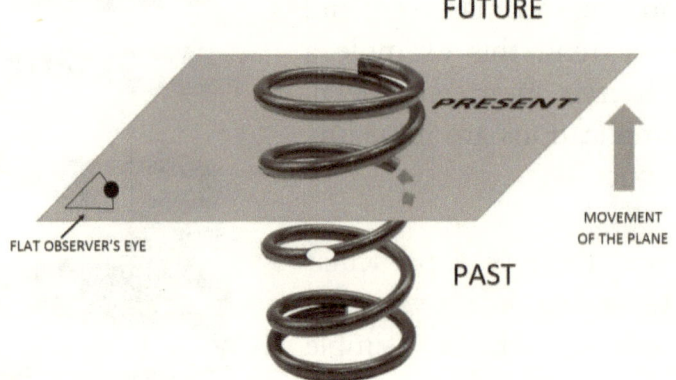

After a little while, that variation is no longer visible and the disruption no longer exists, it is finished, it belongs to the past, an inaccessible past. And yet, we can see very well that the discontinuity on the spring is still there and continues to exist. In fact, we see the spring in its entirety in its three dimensions. In other words, if the flat character is unable to perceive the true nature of the three-dimensional object, we continue to see the cause of what he has seen in his past, but also those events that might take place in his future."

Jan followed the old man's example attentively and asked: "Then you are saying that it is licit to think that if a being were living in a world with more than three dimensions, he could be able to see both our past and our future existing simultaneously in another area of his space. Isn't that so?"

"Exactly. However, you must allow for the fact that this is just a rough example and I am using it only to facilitate your understanding of the question. There are other renowned examples that attempt to explain in an elementary way the limitations imposed by the number of dimensions. One of the best known is attributed to P.D. Ouspensky. In his book entitled *Tertium Organum*, he points out that a bi-dimensional being whose flat plane is intersected by a coin and by a candle of equal diameter, will be able to perceive only the circles of the respective intersections, which to him will appear as two identical objects."

"Obviously" said Jan as he tried to picture that image in his mind.

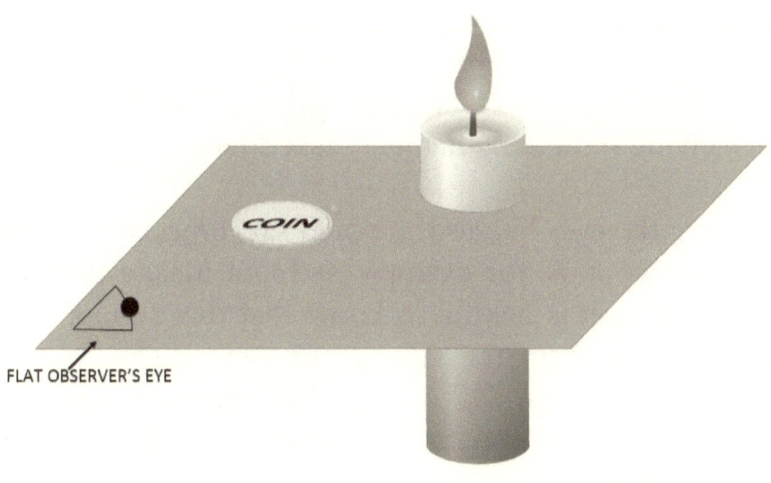

FLAT OBSERVER'S EYE

The old man carried on: "From a view-point that is so far away from the actual nature of the objects, it is absolutely unthinkable that he can conceive how different these are and what their respective functions may be, thus they will remain infinitely distant from his world and unreachable. Similarly, let us think of the enormous evolution we would have to accomplish in order to understand the nature and functionality of objects with more than three dimensions, given that we cannot even see a complete sphere, but only its surface."

Jan observed: "Then it is not just a problem of perceiving an object with more dimensions, but also the fact of being far from being able to understand it. Who knows how infinitely vast is that what escapes our knowledge."

"Absolutely. It is from these very assumptions that, at the beginning of the 20th century, a group of painters gave life to an art movement called cubism, in which they tried to represent how three-dimensional objects and beings would have appeared if observed from a higher dimension."

"Does that mean they were attempting to get our eyes used to perceiving something higher?" asked Jan.

"Not only the eyes but also, I would say, especially the mind. In fact, if we look for example at Dora Maar's portrait by Pablo Picasso, it is helpful to know that that strange human face was born from the attempt to view it simultaneously from all the angles. Of course this approach of viewing things appears very unnatural to us as we are used to seeing only the surfaces."

Dora Maar's Portrait by Picasso.
Cubism was strongly influenced by the attempt to see objects in four
dimensions. If we were able to observe a human face in this way we could
see all its angles simultaneously.

"I am not an art expert and I never imagined this type of artistic expression had a similar scope, in fact I used to consider those strange images only as something quite bizarre" affirmed Jan, surprised.

"You are certainly not alone. In this regard, there is an anecdote about a man who approached Picasso during a train journey and asked him why he did not paint people the way they look, but distorting appearances and proportions. Picasso was not perturbed and asked the man whether he had a photo of his family to show him. When he did so, Picasso asked: 'Is your wife really so small and so flat?'"

"Well, thinking of the analogy with the punishment given by God to Adam, which led to the latter being thrown out from the Garden of Eden and to his confinement in a three-dimensional world, we have to say that life with a low number of dimensions really is a punishment", Jan sentenced with a smile.

"Indeed. And what you call punishment doesn't even spare science because, as we shall see, it needs to strive hard to create very complicated models and equations to explain something that is only an illusion. In fact, recent science has widely proved that the laws of physics that describe the phenomena that surround us become more and more simple and general if they are expressed in a higher number of dimensions. In other words, it is very complicated for a bi-dimensional being to produce laws of physics that explain a phenomenon that is actually simple if analyzed in three dimensions."

The old man carried on: "Let us go back to the flat man who lives on a flat plane and needs to move from one point to the another in his world in the most direct way possible. He believes he is moving on a straight line, which joins point A to point B. He also believes to be better off than his friend who has chosen to proceed via point C, which is not situated on the direct line which connects the point of departure to that of arrival. However, as he walks along the straight line from A to B, something unexpected happens.

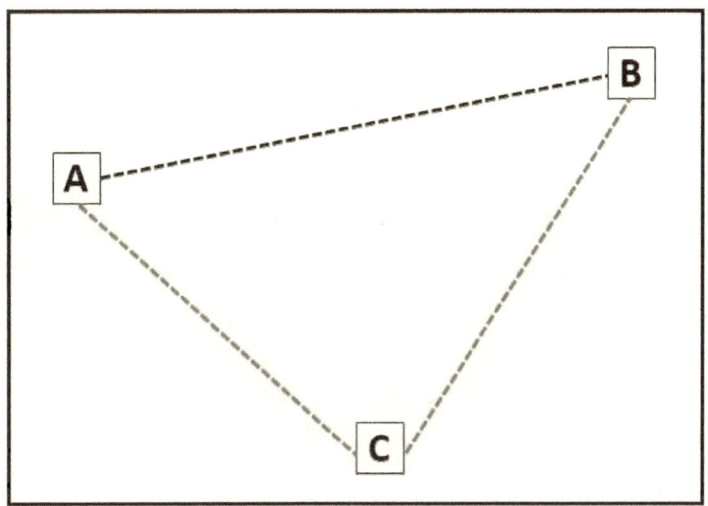

In fact, he begins to sense an inexplicable resistance to his movement and he is forced to slow down and progressively increase his efforts until half-way through his journey, his fatigue lessens and he begins to feel that walking is getting less and less difficult until gradually he begins to feel a push that drives him with such increasing force towards the point of destination that he needs to dig in his heels and break when he reaches it."

"A very strange effect indeed, and what do we owe it to?" asked Jan with interest.

"Be patient, we are going to find out very soon. So, to deal with this unexpected phenomenon, our flat traveller will try in many ways to define a law to explain this anomaly. For example, he could speculate that in some places or in certain

moments, when one begins his walk from one place to another, both the point of departure and the point of arrival exercise on the walker a sort of 'force of attraction' whose intensity depends in some way on the distance from the points in question.
In fact, when the attraction exercised by the point of destination becomes stronger than that of the point of departure, he is attracted towards the final goal and his pace is facilitated."

"True. From his point of view, this would explain everything. It reminds us of what happened in our world when Aristotle, who could not deny that a body's motion accelerates during its fall, invented a principle stating that a body in free fall proceeds increasing its joy, therefore accelerating, as it sees its resting position getting closer." Jan observed.

"Yes, but strange things do not end here. The flat being still needs to understand why his friend who passed by point C did not experience the same phenomenon as he did, and although he did not follow a straight path, he reached destination earlier than himself."

"True, quite frustrating from his point of view." Jan said.

The old man continued: "Now let us leave the flat man to deal with his complicated physics and from our privileged three-dimensional point of view let us look at what has really happened." Taking a handkerchief from his pocket, he put a finger underneath it pushing it upwards into a curve. "See how everything appears simpler? The poor man cannot realize he has climbed an actual mountain which rises between the two points."

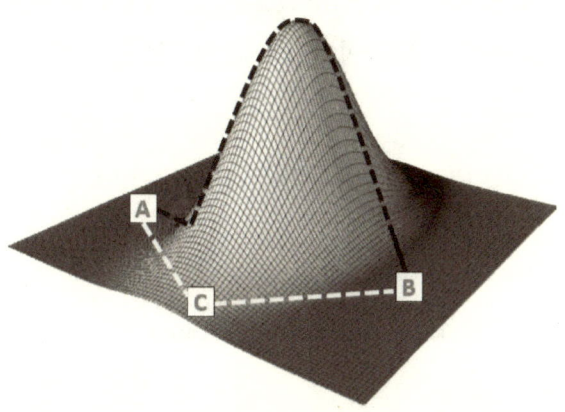

"This explains all !" exclaimed Jan. "Obviously he cannot perceive the mountain because this is located in another dimension which is not accessible to him, but what happened does not look at all strange to us, it is rather obvious."

"You are right. And we know that, each time there is a curvature, the object's motion is explained by the law of gravity. We are therefore led to ascribe to the law of gravity the perceptions and the events experienced by the flat man who climbs the mountain, but the law is itself an illusory force", said the old man.

"Evidently, there seems to exist a relationship between what we perceive and the number of dimensions we live in. This represents quite a problem as it bears significant effects on the complexity of science. Is it possible to attempt to formulate a general principle for this?" asked Jan.

"Well, we could try to put together the two fundamental binomials: Energy-Matter and Space-Time, assuming that as the number of dimensions increases our perception of energy grows while our perception of matter decreases; just as our perception of space increases at the expense of time. This is just a guiding principle, with no pretense of scientific rigor."

"I see, it seems to be a good synthesis", said Jan.

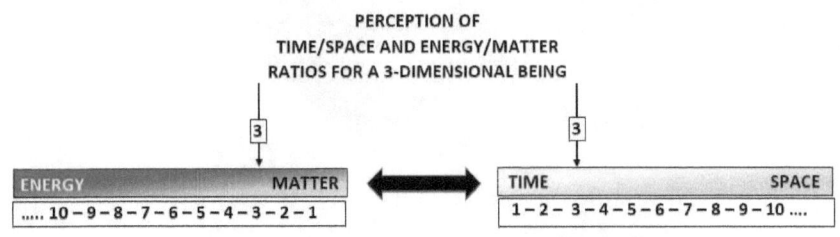

"Now let us move a step forward and get back to our poor flat friends who are dealing with a world that is curved into dimensions that they cannot perceive." said the old man. He carried on:

"There is the possibility that the space of those poor beings is not simply curved into a dimension they cannot perceive, but is actually 'creased' into that dimension, just like this piece of paper." And as he spoke, he crumpled a sheet of paper in his hand.

"Poor things!" Jan exclaimed, "Imagine what a nightmare it must be to have to develop laws that explain such a wide range of strange phenomena ensuing from that."

"Yes, really unlucky. In fact the flat man will be unaware of these 'creases' in space and for him the world will remain absolutely flat. However, as he walks from one point to another, he will be subjected to strange forces that will cause him to accelerate or change direction making it practically impossible for him to walk a straight path.

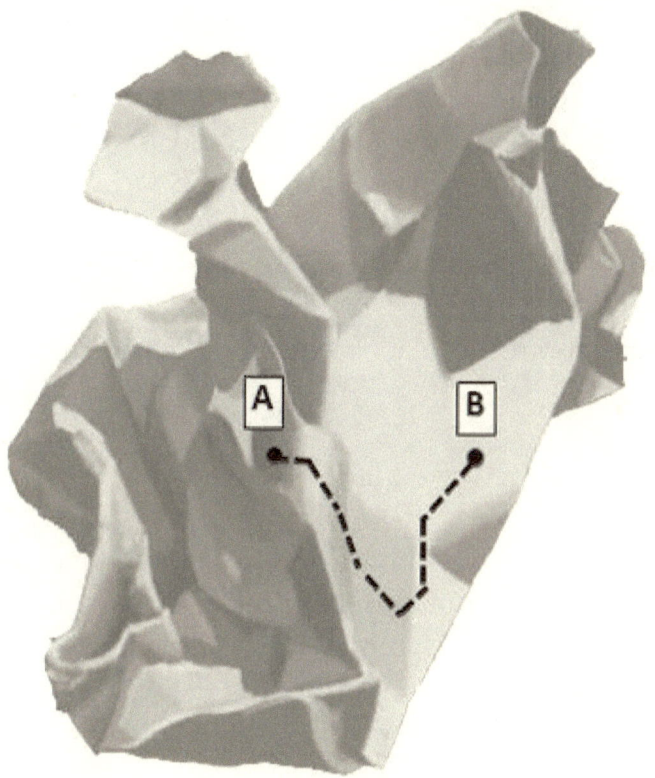

Therefore, in the effort to develop a law that explains his own simple movement, he will need to introduce the action of forces that he sees as real, but which we know to be only apparent. Hence, the laws of his physics that describe movement will need to be very complex and will have to contemplate the case of rectilinear motion as a particular one in which there are certain forces that neutralize each other. In our three-dimensional world it is easier to describe his path."

Jan nodded and said, smiling, "So, in a nutshell we could say that three dimensions are not so few after all, and that it could have been even worse for poor Adam."

"That's right" the old man said, and continued: "Besides, it seems that the limitation with which we have been condemned to perceive the universe could have been even stricter."

"What limitation are you talking about?" asked Jan.

"I am talking about the Planck constant." answered the old man.

"What do you mean?" asked Jan.

"Well, according to Heisenberg's Uncertainty Principle and to the fact that we cannot measure both the position and the speed of a particle at the same time, then, the product of the two uncertainties can never be inferior to a fixed quantity that is valid across the whole of our universe, and this number is known as the Planck constant."

"So it is a sort of 'guaranteed minimum of authentic perception that the good Lord has granted us, isn't it?" said Jan with a smile.

"Yes, a believer may well see it like that" confirmed the old man and went on: "As far as we are concerned, we might ask ourselves how many forces that seem to work in nature and for which we have developed complex laws of physics are only apparent forces, as they are actually due to the 'creasing' of our space into a higher dimension. These forces would disappear if we were able to observe our universe from a higher number of dimensions."

"Is this creasing of our space just a speculation or has this condition been proven with evidence?" asked Jan.

"These are not conjectures but facts that have been experimentally demonstrated, and we can state that we too, like our poor flat man, live in a space that is creased, but we do not realize it, and we strive to produce complex equations and physical laws to describe the behavior of the things around us."

"So, after all we engage in enormous efforts for simple illusions." added Jan.

"Exactly" went on the old man. "In fact, if we take the example of a comet passing near a planet, we observe that its movement is subject to a deviation. Thanks to Newton, we have derived that this is due to a force that tends to attract masses towards each other. The physical law that we have consequently developed also allows us to calculate and predict the trajectory of the comet with good precision.

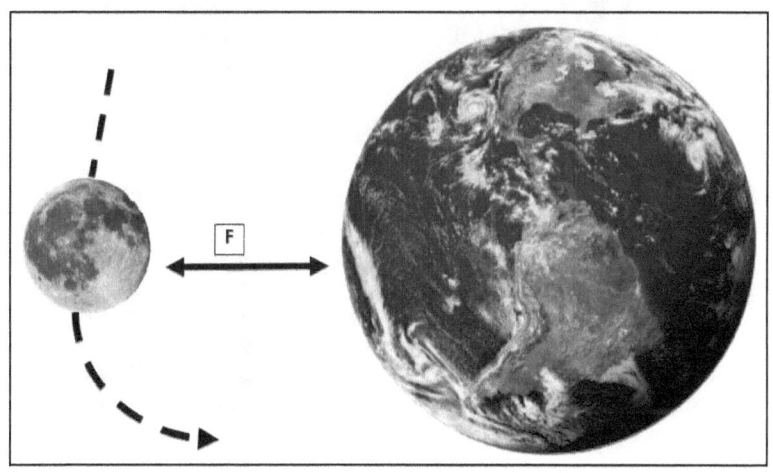

Also, gravitation laws have generally allowed us to calculate the movement of celestial bodies and to send satellites that occupy specific orbits around the earth."

"So how is it possible to state that all this is untrue?" asked Jan.

"We cannot say this is not true, but maybe that the cause is different, and the existence of this force is just an illusion."

"Then what is the real reason that causes the comet to divert from its route?" asked Jan.

"The deviation of the comet is due to the curvature that the planet's mass provokes in the space that surrounds it as described by Einstein's famous equation, which states that the Matter-Energy magnitude causes the curvature of Space-Time."

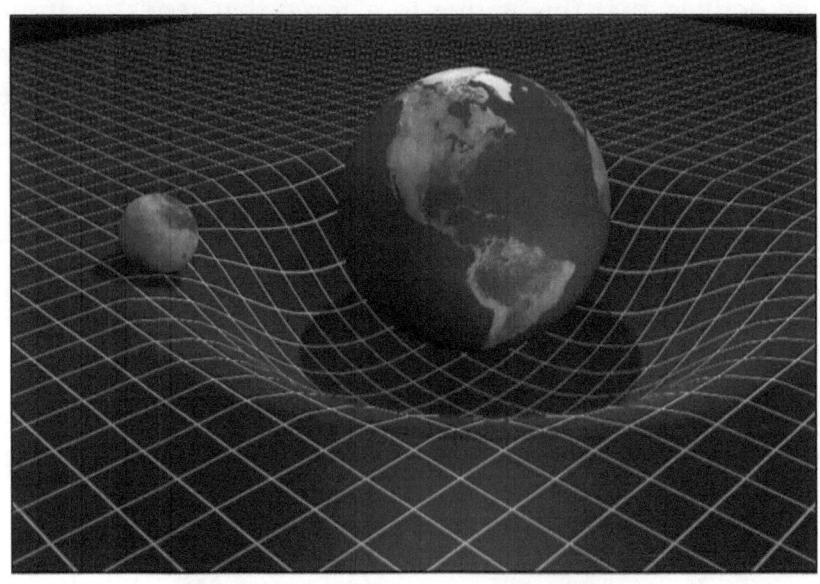

"So does this mean that a big mass, like that of a huge planet, not only alters space but the flow of time as well?" asked Jan.

"Of course, none of these magnitudes can be subject to a variation without the other being subject to a corresponding variation"; answered the old man.

"Yes, I understand." said Jan, almost surprised to be able to so easily grasp and visualize something that would have appeared almost incomprehensible until just a few hours earlier.

The professor went on: "But there is more. You actually need to think of space and time as two different and inversely dependent measures of a single entity. It's almost as if we were to measure the level of water in a glass starting from its edge and from its base. The two magnitudes are inversely dependent but we are measuring just one entity. After all, Feynman himself (Nobel prize for physics in 1965) in his renown essay *Six not-so easy pieces* written in 1994 literally writes: '*... nature is telling us that time and space are equivalent; time becomes space; they should be measured in the same units.*' A similar comparison can be carried out referring to the matter-energy binomial."

"Then, if I understand well, the fundamental magnitudes are not four but two: one is called space-time and the other energy-matter."

"Let's say that, for the moment, to think of and to observe the world according to this assumption helps us to better understand what really happens around us, but I would not exclude further steps forward with the integration of fundamental forces in the universe."

"So you mean that there is no limit to the "creasing" of our world into higher dimensions?" asked Jan.

"It is even worse than that." said the old man with a grin. "We can speculate that not only is our space 'creased' but also 'pierced' here and there. For example, when space is curved by a large quantity of energy-matter such as the mass of a big planet or of a star, light is diverted (as it has been experimentally demonstrated during a solar eclipse when the position of some stars resulted slightly different from that which could be observed at night).

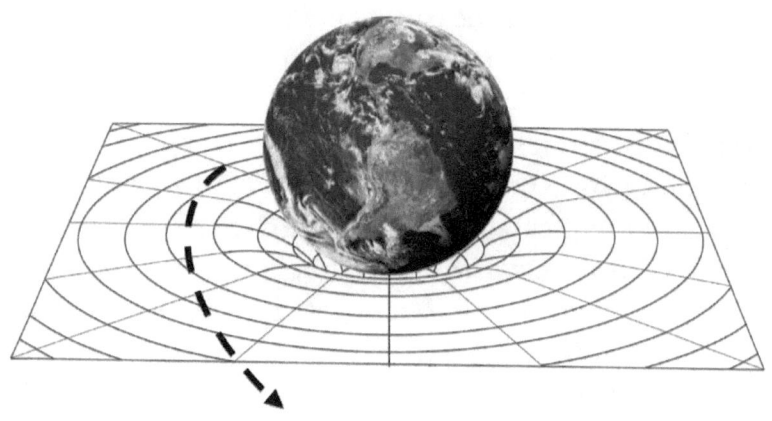

But when the density of the object that produces the curvature grows enormously, like in the case of a dying star, the distortion of space is so significant it produces a real "hole" in the fabric of space-time, something that astronomers call a black hole."

The old man kept crumpling the paper sheet he had used during his demonstrations, piercing it repeatedly, then he carried on: "When this happens light is not just diverted, it is literally swallowed up by the hole. That's why it is called a black hole: because we can't see it."

"Fascinating" - said Jan - "then these incredible anomalies could be anywhere, even very close to us, without us being able to see them or be aware of their existence."

"Certainly, but the most interesting fact is that as we get closer to the hole, reality begins to change more and more, and balance between space-time and energy-matter, which is otherwise fixed by the number of our dimensions, starts to fail, producing strange and unexpected phenomena.

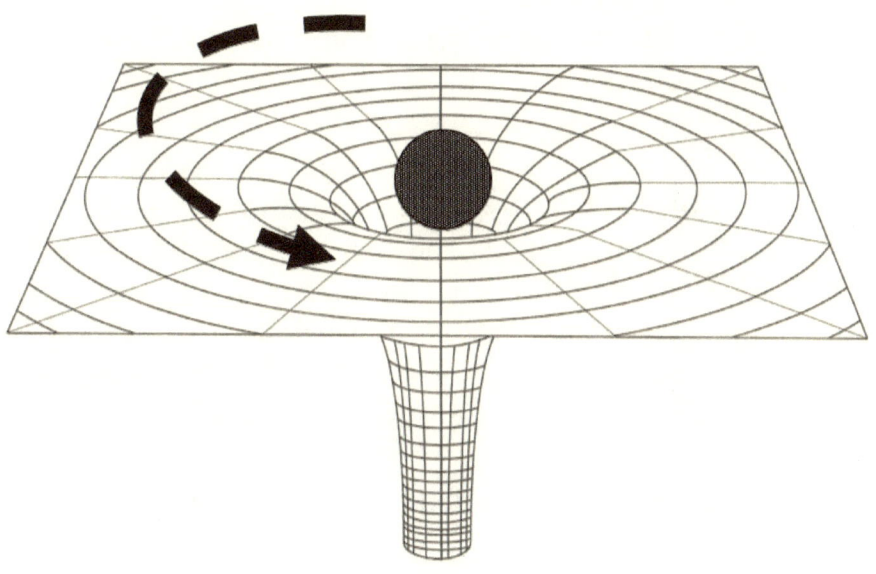

To better understand this we need to get back, as usual, to our poor flat man who lives on a paper sheet and who, unaware of the creasing, keeps producing complex physical laws to explain the "strange forces" he experiences along his path.

If by chance he happened to get close to one of the holes existing in his world, he would probably think he was going mad as he observed and experienced phenomena he could not understand before being invariably sucked in. What happens to him later, should he survive, would be interesting to find out. What is certain is that he will have permanently abandoned his world for another one in which the observable laws and phenomena could be very different from those he was used to."

"So there are huge areas in space where the laws of physics we know no longer work?" asked Jan.

"Yes, but they are not necessarily huge. It's true that when we refer to black holes we are used to thinking of them as the effect of enormously big and dense masses, like those of a dying star that is shrinking, but theoretically, any mass density (independently from its dimensions) that is able to curve space-time around it sufficiently to be able to capture light can be considered the generator of a black hole."

"Then could very small black holes exist?" asked Jan.

"Maybe, but perhaps they would not remain small for long, since they can swallow up anything around them. So we can suppose that if they were sufficiently big, they would grow very rapidly." answered the old man.

"And how could we get to know of the possible presence of 'black holes' in the space around us, if by definition they are invisible?" asked Jan.

"I do not believe there are any real black holes close to us, because in that case we would already have ended up inside them and would not be having this conversation now. Nevertheless, there can be places where the curvature of space-time is inferior to that which is necessary for the formation of a black hole but which is at the same time sufficient for us to experience unusual phenomena."

"And how can we identify these anomalous zones?" asked Jan.

"I believe in several ways; for example by observing that in some places the values of some of the magnitudes we know, such as gravity or time, present significant variations in respect to the average values we observe elsewhere" answered the old man.

"But we already know that gravity, like magnetic fields, is not constant and actually varies according to where we are on planet earth. So how can we distinguish 'normal' variations from the 'special' ones in order to conduct further investigation?

"You see, these facts should encourage us to carry out better investigations, but maybe we have neglected them because the variations are very small and practically irrelevant to our lives, from a practical point of view. However, if we think that time flows differently for two people who are respectively located on a mountain and at sea level, this is sufficient to understand that the relevance of this phenomena is significant. And please note, we are not talking about suppositions or of fantasies, but of proven and measurable facts."

"Of course, this is indeed extraordinary because even though the variations are very small, we are used to thinking that time flows in the same way for everybody", said Jan.

"And yet we have been dealing with these variations for quite some time. For example, the global positioning system, known with the acronym GPS, would not function if those who projected it had not taken into account what we just said" Henry explained.

"So these concepts apply to our daily life as well, without any need for the common man to ponder upon what they mean from a conceptual point of view. For some reason we manage to enjoy the technological benefits produced by things and phenomena that we do not see, but as much as the image of ourselves within the universe is concerned, we limit ourselves to not consider what we cannot see nor touch" observed Jan.

"That's right. For instance, time flows slower for a GPS system satellite than it does for us, because it moves at a remarkable speed, at about 3,8 Km per second, that is 13,600 Km per hour. Such expansion of time would cause, if not corrected, an average error of above 2 Km in the process of identification of a point in our journey. This would undermine the preciseness and the very usefulness of the whole system.

"Truly astonishing" observed Jan.

"But even concerning our everyday life on earth, we should stop and reflect on the conceptual bearing of what we observe."

"Yes, maybe we should make sense of the relations that exist between all the aspects." said Jan.

"Good. Let's try then to put together what we know for certain:

- We know that the clock of someone who is in the mountains works differently to that of someone who is at sea level.
- We also know that this variation is related to the diminishing force of gravity that is experienced in higher altitudes.
- We know, at last, that gravity is itself caused by the distortion of space-time and therefore its measurement can tell us to what extent space-time is curved in that point.

So, as you see, even though we have different elements to reflect on, perhaps we are not doing enough to examine the ensuing implications." said Henry.

"This is because the perceptions of our senses satisfy us and we are unaware as to how limited they are" observed Jan.

"Exactly. In fact, even if we are somehow aware of the existence of a world of phenomena that escapes our senses, most of us have learned to accept its existence without letting the concept penetrate our innermost consciousness.

For example, by now we all have accepted as obvious the existence of radio waves even if we cannot see them, but how many of us have changed, deep within ourselves, the awareness that we live immersed in phenomena we do not see?

How many of us can see ourselves, with the eyes of our minds, as beings who are continuously pierced by radio rays, by cosmic rays and who knows what else? Or hit by light rays (infrared or ultraviolet) that we cannot see, or surrounded by forces and phenomena that unfold in other dimensions and whose interactions with our world we perceive, only at times, without fully comprehending their nature?"

"Well, we usually tend to leave this field to charlatans or others in search of celebrity who speak somewhat vaguely, about strange energies, interventions by extraterrestrial beings and other fantasies of the kind." said Jan.

"Instead, it is necessary to consider this issue seriously, along with the certainties that science has given us, and allow these concepts to mature in our minds. We need to practice rethinking ourselves as part of a small world, which is limited by our perceptions, and remind ourselves in every moment that events we consider belonging to the past, are maybe still present, and exist forever in some region of space that we cannot perceive, which is nevertheless real, though not physically reachable."

"This is all very beautiful but there is something I do not understand" said Jan, who added: "If time does not exist but is just a sort of distortion with which we perceive parts of a phenomenon that exceeds our dimensions, how come we perceive time as ever flowing and just in one direction? In other words, why does time never flow backwards, allowing us to go back towards the past?"

"Be careful", warned the old man, "it is not time that flows in one direction, but the whole universe that moves in the direction of expansion. Therefore it is logical that time, which is just a perceptive aberration, appears to us to always move and only in one direction".

Jan listened to the old man's words, every now and then trying to create images in his mind that could represent these deep concepts. For sure, seen from that perspective, his whole life appeared different and endowed with other goals and meanings, whose enormous bearing, Jan felt, he still could not comprehend.

He thought about what it really means that the past has not actually passed but still exists somewhere and, given the recent enormous grief he was undergoing, he recalled the tender image of his mother. He imagined her in her best years, reliving his own memories. He would have given anything to be able to live them again and although he was aware of the impossibility of a similar event, the old man's words had at least enabled him to relive in his mind those lovely and moving moments.

The sun was now setting and sharp pangs of hunger reminded Jan that, immersed in their interesting discussion, they had skipped a meal.

The old man left the hut and after a while came back with a kind of a still steaming pizza which he had kneaded and baked on a flat stone. They opened a can of beans and, having relished their simple meal, withdrew to their respective spaces to sleep.

Jan lay at length, immersed in his thoughts with eyes wide open. Those reflections had presented him with a new view of the world, which, although unsettling on the one hand, on the other had also given him a strange feeling of inner peace. For the first time he felt relief from a sense of anguish that only now – he realised – had pervaded him. He tried to understand what could be the cause of the pleasant feeling, and reached the conclusion that it was the very idea of the nonexistence of time that induced him, curiously, to make sense of the experience of life in a different way. For the first time he reflected with deep introspection on his perception of the path of human existence. The concepts expressed by the old man offered him powerful tools with which to investigate within himself as he had never been able to do before.

He had never been conscious of the fact that it was actually death which, in his subconscious, deprived life of its meaning. To live and accumulate knowledge, experiences and emotions only to see them invariably sink into an abyss without hope, appeared to be not only useless but, in some cases, seemed to mock the human condition.

But now a new perspective unfolded before his eyes. If it was actually possible to see life as something continuous and alive in each and every moment and if death could really be considered just as an instant, although an extreme one, of bodily life, then everything acquired a new and marvelous meaning. The path of human life now appeared to him as nothing more than the exploration of all the multidimensional "sections" of a human being and, although the past remained unattainable and the future unfathomable due to sensorial limitations, everything suddenly appeared to make sense.

"Then why" he wondered, "why are these concepts, this scientific evidence, elaborated in the last century, not taught to young people in their first years of adolescence, even in a simple way? Why not help young people to conceive of themselves in higher terms? Is it ignorance, fear or lack of interest?"

Certainly not, he concluded; but then he was unable to find an explanation for a behavior that prevented young people from reaching such an important level of consciousness, compelling them to re-consider their strongest beliefs in adulthood. And why did the teaching of religion not take advantage of all this? After all, not only did it not affect faith, but to the contrary, for a believer all this obviously represented the greatness of God.

Jan felt sleep approaching, and he yielded to it with a peacefulness he had never known before.

FOURTH DAY

QUANTUM MECHANICS

The old man was woken by the sound of an axe as Jan chopped wood with which to light a fire for the oven.

"Were you not able to sleep?" asked the old man as he stepped out of the hut.

"Quite the opposite. I have slept very well and I am raring to go" answered Jan with a smile. "And I am feeling really hungry".

The old man laughed heartily, and together they prepared the flat breads they baked in the oven, which Jan proceeded to wolf down with some hot milk.

"So, is your idea of the world changing?" asked the old man.

"Completely" answered Jan, "However, although I sense a new way of looking at things, I still have much confusion in my head".

"That's logical" answered the old man. "Of course it's not simple to review the certainties which we thought to be unchangeable ever since we were born and replace them with something we cannot fully understand. However, the idea you had of the world is definitely wrong while the one that is forming in your mind now is surely more coherent with reality, though still not completely clear."

"That is true" agreed Jan.

"The thing is" went on the old man "that mathematics and science generally do not suffer, fortunately, of sensory limitations, and their development has helped us to understand that our world is not as we instinctively perceive it, but it's much more complex and we can't figure it out just by using our senses. Therefore, because some physical laws that appear absurd to us actually work perfectly, so much so that we have already been using them for quite some time in the field of industry, then our perception of the world must be wrong and we need to come to terms with a bizarre reality which holds true some assumptions that look unrealistic to us."

"Are you talking about conjectures and theories or about something with scientific evidence?" asked Jan.

"Totally proven in experimental reality" answered the old man, who carried on: "You see, classical physics, Newton's physics to be clear, is based on the perception of the surrounding world according to our senses. In other words, that physics has produced laws that explain why the phenomena we observe do occur in the way we see them. However, in the beginning of the twentieth century we realized that this was no longer sufficient to explain many phenomena. Therefore, scientists needed to develop a theory that had to be independent from our perceptions and from common sense. This is quantum mechanics. It analyzes consequences that appear absurd to us, although they are totally true and proved."

"Could you give me a couple of examples?" asked Jan.

"Certainly. What would you think if I told you that, under particular circumstances, the consequences of an event can precede their own causes? Or that an object can simultaneously be located in more than one place? Or that if you open a drawer, look at what is inside and then close it again, you cannot be sure it still contains the same things before the opening, during the observation and after it was closed?" asked Henry.

"Crazy," said Jan with eyes wide open and carried on, "and this is the kind of thing that has been experimentally demonstrated?"

"In a way yes" answered the old man. "And now we shall talk about it more in depth."

"I am too curious and, before you start talking about the wider frame I would like you to describe what kind of experiment could provide evidence for the example regarding the content of the drawer?" asked Jan, with an almost challenging tone.

"Ok, as you wish" conceded the old man, who went on: "Imagine carrying out the following experiment in a big box:

we throw a succession of very small, identical particles against a special kind of metal foil.

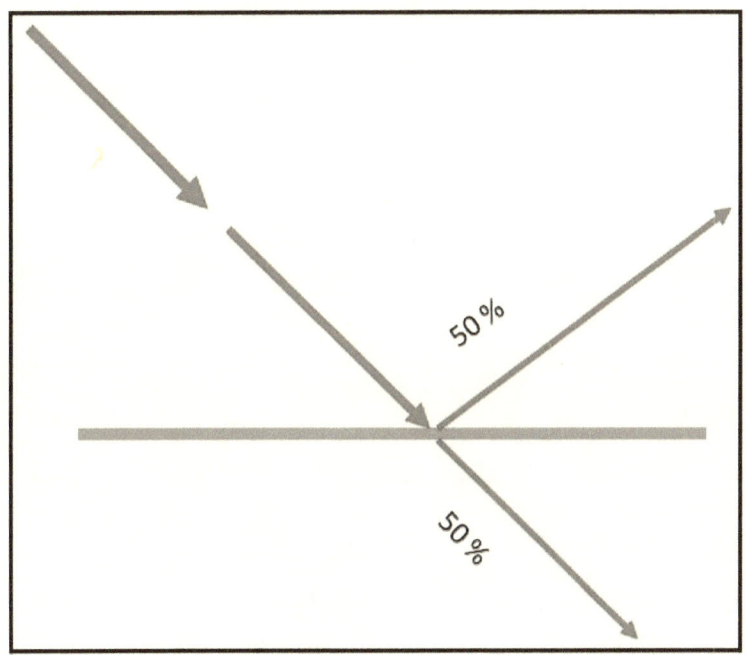

Now, two things can happen to each particle: it either bounces against the metal foil or it goes straight through it. It is possible to produce an adequately thick foil so that 50% of the particles go through it and the remaining 50% bounce back."

"Ok" said Jan

"Good, now suppose you insert a thicker obstacle across the bouncing trajectory, which cannot be crossed, and add another metal foil that is as thin as the previous one and which can more or less be crossed by 50% of the particles.

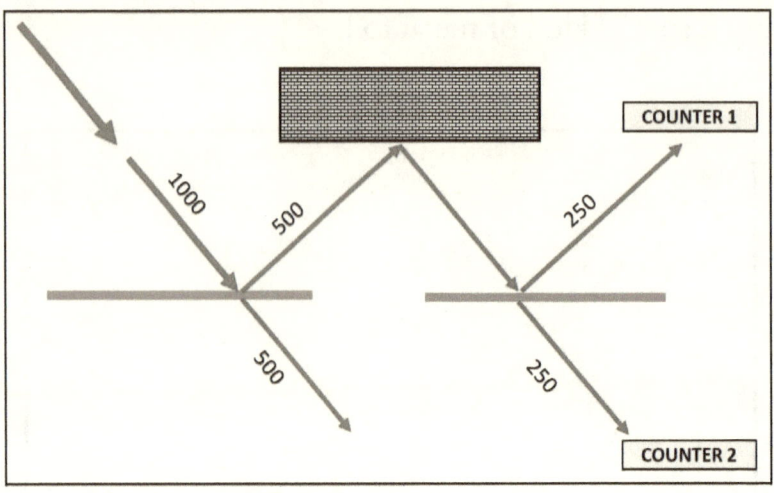

All the 500 particles that reach the wall bounce back and, considering that 1,000 particles were launched, you should be able to count 250 particles crossing the second foil and other 250 bouncing back. If you have positioned two automatic counters at the right points, you should be able to measure exactly 250 particles on each side of the second foil."

"Well, everything as expected then" objected Jan.

"So far, but now comes the interesting part. Imagine placing a second wall on the path of the ray of particles that has crossed the first foil. At this point you should have respectively 250 more particles crossing the second foil and as many bouncing back. Then your automatic detectors should measure 500 particles each."

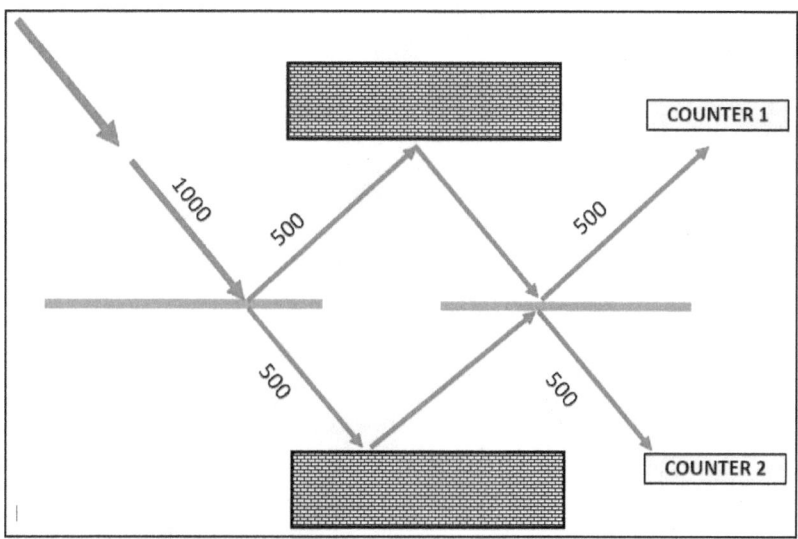

"And that isn't so?" asked Jan.

"Right. Now, for some mysterious reason, all the particles are going to end on counter 1, which points to 1000, whereas counter 2 points to 0."

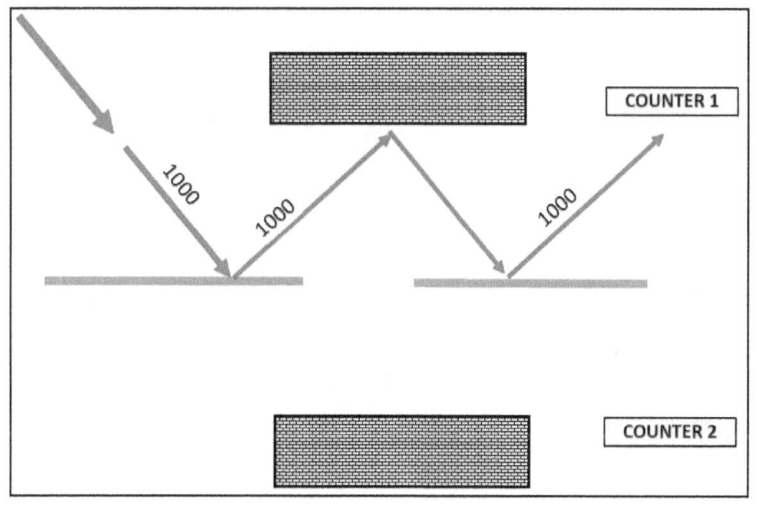

"Incredible" said Jan

"And we are just at the beginning. We can now observe some strange facts. The presence of the second wall actually affects the behavior of the entire flow of particles. They are no longer going to cross the foils but they are all going to bounce back. Therefore, counter 2 will detect none.
Now we should ask ourselves how is it possible that the entire flow of particles, which bounces back when the second wall is not there, notices that a second wall has been inserted. We do not know, but we can measure that this takes place and also that it would take place even should we position the second wall at a great distance from the first. It's as if the system and the second wall are bound by a mysterious connection that is not affected by distance."

"I do not understand." said Jan.

"In fact it appears absurd. And yet it's what happens and the only way to explain it is that each particle is simultaneously in two different places, that is, above and below the foil, and that it can thus know if the second wall is there or not, and consequently vary its behavior" said Henry.

"So, this is one of the strange facts you mentioned a moment ago, an object which is in two different places at the same time" said Jan.

"Indeed. But now things get even stranger: if I do not limit myself to just registering the counters' data, but also observe the system in order to understand what has happened, well, everything goes back to normal and I will find I have 500 particles on each counter."

"You mean that the system changes its behavior depending on whether you watch it or not?" asked Jan.

"That's right. We do not really understand why, but I can assure you that this is exactly what happens. So the real difficulty is that, unlike in classical physics, we need to work with magnitudes and phenomena that we cannot observe, but only try and imagine."

"Right, but it really looks like nature is very different from what we perceive." said Jan.

"Yes, and it has been some years since we started to apply the laws and principles we have been able to derive, even although we cannot properly describe which concepts we are dealing with and using. We are happy to use devices that have now become common such as lasers or smart-phones, but we would never have never been able to produce them if we had not been able to develop and apply those quantum mechanics equations whose results appear so implausible. All the progress achieved in developing the most modern electronic devices is due to quantum mechanics and to the laws that regulate the quantum-bit (or qbit).

The utilization of the quantum approach in the design of electronic circuits and the passage from the simple bit to the qbit has allowed us to improve the performance of our most common devices pushing it to levels that were absolutely unreachable with previous technologies. Well, we have been able to grasp these laws, we know that they also exist in our world, but with our common understanding we are unable to conceive how this can happen. How could we ever imagine, in fact, that an entity, for example an electron, can be found in two different places at the same time? And yet that's exactly the way it is."

"It's as if we were exploring a pitch dark environment, gaining awareness of its possible pathways only through touch, without being able to see them" affirmed Jan.

"Indeed. The conclusions we have reached through quantum mechanics are so surprising and strange, they led Niels Bohr to pronounce a sentence that has become famous: ***Anyone who is not shocked by quantum theory has not understood it.***"

"Very funny, but it does convey the idea" said Jan, laughing heartily.

The old man went on: "And the funniest thing of all is that nature seems to enjoy making fun of us, because the effects of quantum theory take place provided we do not observe them.
In fact, if in a given moment we decide to observe the quantum system, it stops functioning according to its laws, which for us are "transcendental" and starts functioning in a way that is compatible with our perceptions. In a way it's as though our simple observation reverts the system to a more limited universe where these phenomena do not take place."

"This explains the example of the drawer which you mentioned earlier." said Jan.

"Yes, even Einstein could not easily accept this, albeit he could not deny its reality. In fact, during a scientific debate he had with Niels Bohr he stated: '***I like to think that the moon is there even if I am not looking at it***'.

In other words, we have learnt how to make use of something we cannot completely understand nor concretely imagine. We produce devices such as smart-phones, computers, cameras that work thanks to equations and principles which originate in a world or dimension that is not at all clear to us. We are like the flat man who has learnt how to make use of the candle, maybe to get warm, without however being able to see it nor understand its functions." concluded the professor.

"Has any of this got to do with the problem of time and dimensions?" asked Jan.

"Each theory or point of view that attempts to interpret the universe that surrounds us needs to be verified, analyzing points of accordance and discrepancies with respect to the other theories. In fact, one of the major efforts on the part of scientists in recent times has been to render the theory of relativity compatible with quantum mechanics".

"And what results has this effort yielded?"

"Do not overestimate me, my boy. Things are complicated for me too and I don't know much about this. However being acquainted at least with the most basic principles can help us to use our creativity and to grasp a few connections".

"For example?"

"Well, as we partly saw in the example of the two metal foils, one of the strangest peculiarities demonstrated by quantum physics goes under the name of '***entanglement***'. This means that when two particles (or even objects) become 'entangled', their properties and to a certain extent their destinies, are connected to each other in a way that is mysterious to us. In fact, even if we were able to physically separate these entities, taking them to the extreme opposite sides of the universe depriving both of them of the possibility of 'knowing' or of 'seeing' the other, if either particle is subject to a variation of its current state, a correspondent variation will occur in the other at the same time."

"A bit like in the example of the foils, regarding the strange connection between the first and the second wall" commented Jan.

"Exactly. Now if we consider the behavior of electrons, we know that every electron has a 'spin', that is a completely random rotation in any direction, clockwise or counterclockwise, until we observe it. In the very moment in which we observe it, the direction of its rotation is affected in one way or another. Hence, if this electron is 'entangled' with another one that has been taken, for instance, to the moon, the observation of the former will most certainly bear an effect on the latter, causing it to rotate in the opposite direction. This fact has been scientifically proven through numerous experiments, but how it happens is incomprehensible to us."

"And so?" asked Jan.

"Well, to answer your question and to show you how further considerations on time and dimensions can suggest some interpretations regarding these facts, let's try to consider the entanglement of the electrons from another point of view. If we are not surprised by the fact that two particles mutually affect each other's behavior at the beginning of the experiment, because they are close to one another and are therefore able to modify the condition of their partner, why are we surprised that this connection still exists when the two elements are separated by great distance?" asked Henry.

"Well, the fact that the enormous distance should not allow any interaction between them" answered Jan.

"Yes, but this is what we think because instinctively, we take for granted that the initial moment of the experiment is terminated. In other words, we continue to fall victims of the deceit of time".

"Uh... the usual story..." said Jan, fully grasping the direction of the discussion.

"Yes. Now if you try to think that the initial moment of the experiment has not ended but still exists somewhere in space-time, it is obvious that the particles continue to be connected even if they are separated in another region of space-time".

"So you are saying that the starting point of the experiment, that is when the electrons are close to each other, and the following moment, when they are separated by an enormous distance... that these moments are actually ever present and therefore co-exist in space-time?" asked Jan.

"Exactly. So the fact that the connection persists appears to be more acceptable. In other words, the property of 'entanglement' does nothing but remind us, yet again, that time does not exist."

"Right, this is a further confirmation", said Jan.

"And it's not the only one. If you remember, a while ago I was telling you that the consequences of an event can precede the event itself." said the old man.

"Yes, something else that is totally incomprehensible" answered Jan.

"Well, you see, for quantum mechanics this is less strange than it may appear to us. Whilst we consider an event that has taken place and has ended to be unchangeable, certain and determined, just like a series of potential photographs that we may have taken during the course of the event, for quantum physics it is not at all determined, provided we do not observe it. The past, just like the future, exists only in terms of the probability that one of the possible permitted events has taken place. Therefore, any observation on our part of an event in the present, not only can affect its future, but, in a certain sense, also its past. Please note again, that we are not talking about anything abstract. What we have just said has been experimentally demonstrated on numerous occasions. In this regard, the most famous example is the so-called '*delayed-choice*' experiment, conceived by J.A. Wheeler in 1978 and performed by Carrol Alley in 1984."

"And has this experiment actually demonstrated that a modification of the present can affect its past?" asked Jan.

"Something very similar. The experiment is rather complex to describe and it is limited to the very small dimensions of photons, but you can be assured that it has demonstrated that by performing a certain action in the present can modify the past, so as to make one of the possible pasts more or less probable with respect to the other ones" said the old man.

"So in a sense the past is still alive and not frozen in history. Isn't this the evidence that time does not exist?" asked Jan.

"Excellent observation!" exclaimed the old man aloud.
"You are slowly acquiring the right mindset to observe reality for what it actually is, in as much as it can appear strange to our common sense".

"Yes. The more I learn the more it appears less strange and much more beautiful." said Jan. Suddenly his attention was drawn to an unusual, coarsely framed photograph, standing amongst the others on the table. The photograph showed a row of human-shaped wooden legs, side by side and suspended in what appeared to be the production line in a factory of prostheses for people with disabilities.

"Why did you choose to frame this photograph?" asked Jan, and added: "This is very unusual, does it symbolize the journey of man?"

"No" answered the old man, "even if your supposition is a very nice one. That photo struck me in a particular way at the time in which I was beginning to pursue knowledge, which is the reason I wished to keep it."

"And what was it about this image that struck you so much?" asked Jan.

"You see" continued the old man, "that photo made me reflect on the fact that once an object is created, one of those wooden legs for example, it 'exists' in the multidimensional world, of which we see only a part. As soon as each of those legs is created it becomes the replacement for a physical leg. Obviously, we still do not exactly know whose leg it is going to replace and we can imagine that the latter is still part of the body of the poor man who is going to lose it".

"That is a little macabre but nevertheless exact." commented Jan.

"As you see, we are talking about time, it seems we cannot avoid it, but this is not really the way things are. In one of the possible ramifications of his life, Mr. X is, for example, in Seattle. He has lost a leg, which has been replaced with the third prosthesis in this photo. Because time does not exist and any possible ramification of events already exists, that means that in the very moment in which the prosthesis is created its simple existence modifies an infinite series of possible paths, creating a new one that gains probability of taking place. A probability that increases when, for example, that specific prosthesis is sold by the manufacturer to the hospital in Seattle. Ultimately, through a great number of other factors that operate in that direction, that particular reality further increases its probabilities of occurring, until it actually takes place. All this applies to any existing person or thing".

Jan pointed out: "So you mean to say that whenever something exists or is created, it lives in a multidimensional world and generates an infinite number of ramifications or sub-ramifications of possible developments for particular events. Is that correct?"

"Exactly" confirmed the old man.

Another day passed by and Jan was now beginning to get used to the constant amazement that the words of the crazy old man aroused in him. Through simple objects or examples, a new vision of the world was beginning to take shape in his mind, and it was so different from common understanding.

After eating their usual frugal dinner they retired for a night of rest while sporadic gusts of wind could be heard hissing through the many cracks in the walls of the hut.

FIFTH DAY

KNOWLEDGE AND SELF-KNOWLEDGE

By now, Jan was feeling quite different. Those long days spent with the old man and those glimpses of reality that had unfolded before his eyes were changing him for good. He was beginning to sense his place in the world in a different, much more complete, way. At the same time he was aware that the understanding of many aspects of this new reality would remain inaccessible to him forever.

After having eaten the usual breakfast with the old man, he asked, "The perception of myself and of the world around me is changing rapidly thanks to our reflections. Now I am curious to hear how you see yourself."

"You see," the old man said, "for a man, knowledge of oneself is one of the most difficult exercises. Precisely for the same reason why a pencil cannot possibly write on itself, a man cannot directly perceive himself. In fact, self-knowledge is always achieved through comparison and not through direct observation; therefore it is much more difficult and deceptive.

Furthermore, knowledge and self-knowledge go hand in hand. You cannot say that you know yourself without knowing the world of which you are part."

"Yes, it is a difficult exercise, especially as the real world is far greater than what we can perceive with our senses. So much, I well understand by now." said Jan.

"That is right. And how would you attempt to explain to the flat man that there is a greater world?" asked the old man.

Image from the film: Flatland

Well, I would try to explain that there is a dimension that is perpendicular to the first two, along which you can move and see the flat world from above." said Jan.

"But the word 'above' has no meaning for him, what's more he cannot conceive a direction that is perpendicular to the only perpendicularity he is aware of, just as we are not able to imagine a fourth dimension that is perpendicular to the three we know."

"Yes, I do understand it would be difficult for him, but I would ask him to make a mental effort and, although I cannot materially portray it, I would suggest that he accept the idea that it can exist." insisted Jan.

"You mean that you would suggest to him to try and interpret the phenomena that he is able to perceive in this new light?"

"Exactly!" exclaimed Jan.

"Well, he might be able to accept the idea that what he considers as the past and the future are simultaneously in a certain direction of space that is precluded to him, but how will he able to take a step forward in understanding the differences between the coin and the candle?" the professor asked.

"Obviously, he will not. However he will have gained several advantages from of my explanation."

"Like what?" asked the old man, somewhat surprised that Jan was now so much more confident in handling these concepts.

"Even in his flat world he might actually get an idea of how a three-dimensional object is made. If you think, for example, of the three-dimensional evolution of a square into what we call a cube, the flat man might be able to see its intersections across his flat world.

He might therefore draw a small square within a bigger square on his own plane, to represent the projection of a three-dimensional world. Even if the lines that have been drawn have different lengths for him, he will know that the small square and the big square refer to objects that have equal dimensions for a three-dimensional observer, even if he cannot imagine how."

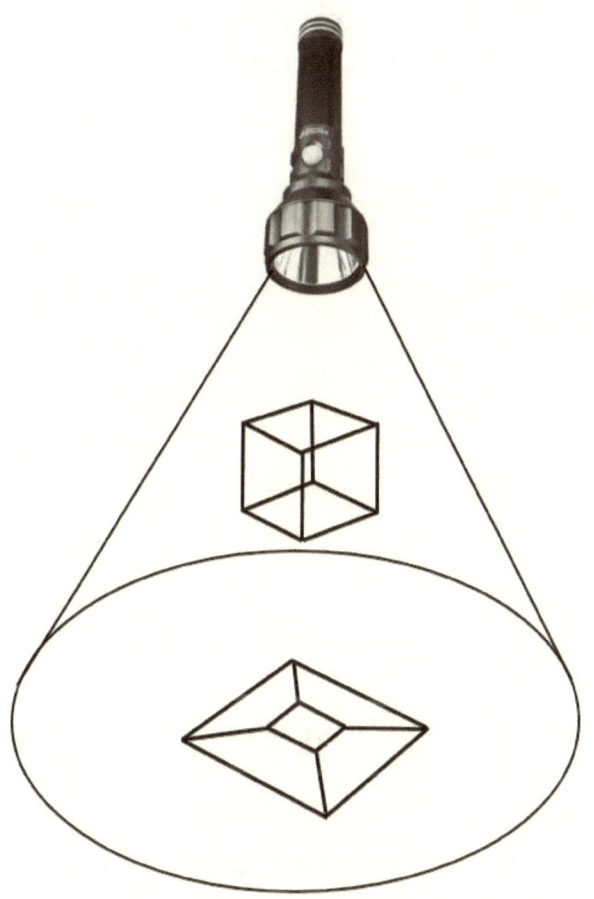

Projection of a Cube in two dimensions

"Excellent!" congratulated the old man. "In fact, starting from this very idea, the English mathematician C.H. Hinton managed to represent a Hypercube, (or Tesseract), that is, a four-dimensional cube.

His representation shows a cube within another cube, which is necessarily larger. Yet we know, even if we cannot graphically portray it, that both cubes are of the same size if seen from a four-dimensional perspective."

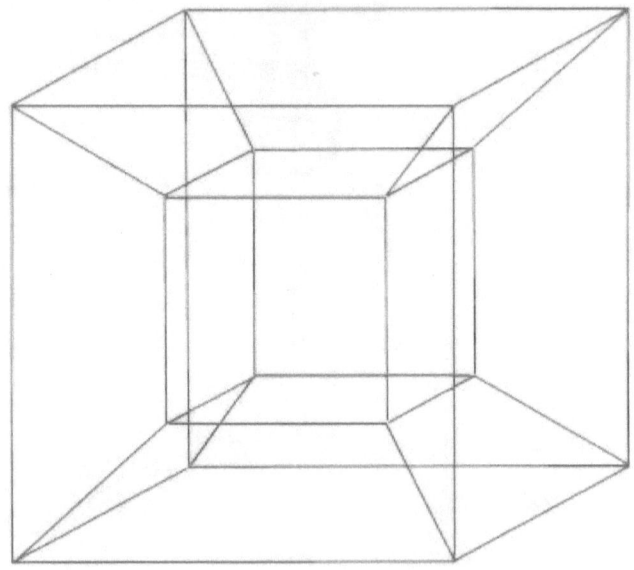

Projection of a Hypercube in four dimensions

Jan went on: "Then he would know that his past, even if it remains inaccessible to him, is not finished, it is still there and living, and his life stretches into an eternal present. In addition, he would learn not to judge phenomena according to how they appear to him but to ask himself whether they serve a higher purpose or function."

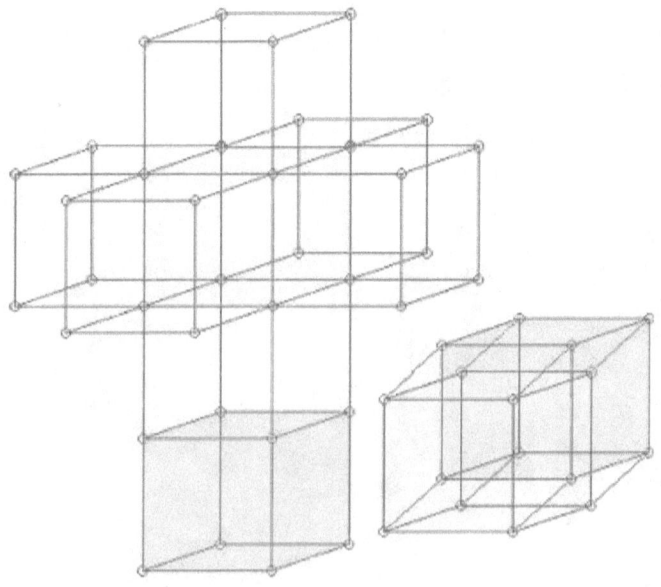

Hypercubes

"Be careful not to let what man does not understand slip into a sort of religion based on beings with higher purposes and goals. This is precisely what science must avoid in order to preserve its pragmatic approach that refuses to recur to faith." warned the old man.

"I do understand that" said Jan, "but when I speak of higher purposes or goals I do so in secular terms, I refer to things that we cannot see because of the limitations of our senses, but that nevertheless exist and are equally real."

Salvador Dalì: Christus Hypercubus

"This is definitely the right approach. Let us not consider, for the moment, the religious aspects of this matter, even though it is not hard to imagine how the flat man would perceive someone who is able to see the past and the future at the same time and also see inside all the rooms of his flat house with a single glance. He would instinctively consider him to be a God." said the old man.

Jan nodded: "Of course. Moreover, the fact that someone who lives in three dimensions would have the possibility to see and to have direct access into the body of the flat being, and the latter would feel observed in what he would consider a supernatural way. For example, a 3D surgeon could intervene on his internal organs without having to make any cuts and hence operate without him even realizing, thus accomplish something of a miraculous healing."

3D Surgeon and 2D patient

"Yes, but the flat being would be mistaken to consider him a God. The idea of God has nothing to do with this. In fact, for believers, God is a being with no limitations, whereas for the flat man we would just have fewer limitations than the flat man himself." concluded Henry, without concealing his satisfaction for the level of Jan's considerations.

"I am thinking how precious it would be for us to have a four-dimensional surgeon." Jan said, laughing.

"Undoubtedly so. But now let us go back to the issue of the expansion of knowledge and the ability to go beyond the context in order to reach a higher state of consciousness. As we have seen, for a two-dimensional being the coin and the candle will look the same and he will never be able to understand the differences between these two objects in the world around him."

"Yes, I remember. But are you now referring to the physical differences between the two objects or to the differences of their functions?" asked Jan.

"Both. In fact, the essence of the object is fruit of its form and its function, and both escape him entirely. In fact, we also found that, in order to be able to completely see an object in n dimensions, we need to have $n + 1$ dimensions. In fact, we are able to see the square entirely, something that is not possible for the flat man because he is not able to observe it from above."

"Of course. And I understand that we are only able to see the surface of a solid body, i.e. a three-dimensional one, because in order to see it fully we should be able to observe it 'from the top' of a fourth dimension, something that applies even more to the hypercube" said Jan.

"In fact we only perceive the 'shadows' of solid objects and in order to distinguish between two intersections that look the same we ought to base ourselves on their functions, going beyond the context in which we find ourselves. Obviously this applies to three-dimensional objects because we can understand their functions, but we would not be able to grasp any function of four-dimensional objects."

"And would a four-dimensional object appear solid to us?" asked Jan.

"Only in its intersections, therefore hardly at all. In fact, because the square is actually a solid for the flat man, and he can bump one of its sides, the same cannot happen with the spring. For him, the solidity of a three-dimensional body is distributed over time. However, you must consider the limitations of the example we have used, which only serves to exemplify the concepts but cannot serve as a truly representative model. In fact, the intersections between worlds of different dimensions are dimensionless, just like the dot or the line and are therefore difficult to perceive if we use size as a reference. However, in principle we can say that a being with n dimensions can perceive as solid the border line that marks the intersection of a n+1 dimensional object with the n-dimensional world, whereas the exceeding part slips into time. Similarly, the two-dimensional being can perceive the circle that is produced by the intersection of the candle but not the candle, just as we can perceive the surface of a sphere."

"And what about the function of a possible four-dimensional object?"

"If we rely on our five senses I believe it is impossible to imagine." replied the old man.

"Couldn't we find another way?" said Jan.

"Personally I do not see how. However, some people claim to have, by instinct and in an almost unconscious way, a sort of a special sensitivity and claim to be able to perceive the difference, for example, between a brick that belongs to a school building and another, apparently identical one, that seals a tomb. The scientific world has expressed skepticism regarding the reality of these perceptions, and many experiments conducted in a scientifically controlled environment, have never proved, to the best of my knowledge, that this corresponds to truth."

"This seems obvious to me" said Jan and continued: "moreover no scientist could ever find any measurable difference, in our physics, between the two bricks that are absolutely identical. However, a scientist in a flat world would nurture the same kind of skepticism if someone told him that there is a perceptible difference between the two circles that are generated by the intersections of the coin and the candle. He would smile and proceed to demonstrate how the scientific measurement of the two circles produces the same result and therefore, there is no evidence that proves the difference between the two phenomena."

"Certainly" confirmed the old man.

"However in the example of the two bricks there is something that is not quite clear", said Jan, engrossed in an effort of imagination.

"What do you mean?" asked the old man.

"The two bricks are identical from all points of view as they were both manufactured in the same factory and at the time they were created there was no knowing whether they would be part of a school or of a tomb. So how can we consider them different?"

"As usual, you are not correctly considering the time factor. The future and the past are a timeless *'continuum'*. The fact that a brick alters its function after its creation does not modify things in the slightest. What happens to it in a certain region of space-time inevitably affects its condition in other regions of space-time in which it is located. The moment of its creation is always present and what happens to it in a different region of space-time alters it also in the moment of its creation." said Henry.

"You are right. I still need to assimilate the idea of the non-existence of time." concluded Jan.

"Besides," continued the old man, "whether we perceive the intersections or not, the interaction with n-dimensional worlds most certainly produces some effects on our world and this is very interesting if we think of the structural aspect of these objects."

"What do you mean?" asked Jan.

"Just try to imagine, with the usual analogy, the intersection of a tree with a flat world. In that case, you would see the lines caused by the intersections of the specific branches with the flat world as separate phenomena and you would not be able to understand how they all originate from a single cause. Well, going back to our world, who knows how many phenomena that we consider separate are actually due to a single cause that we cannot see because our perception is limited?"

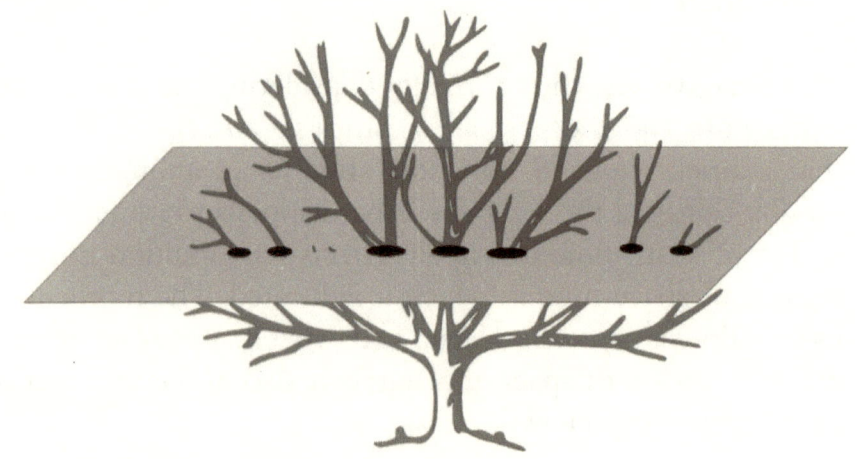

"Right" said Jan, reflecting on the new implications opened by that hypothesis.

"And there are many much deeper issues that need to be considered" objected Henry.

"I can hardly believe it. It seems to me that what we have said so far completely changes the perception of ourselves in the world that surrounds us." said Jan.

"Yet there is much more. Now I ask you to consider something: What is the difference between a living cell and a cell that has just died? Or, for a man who lies dying on his bed, what is the difference between the moment before and after his death?" asked the old man.

"That his heart has stopped beating?" Jan said tentatively, slightly puzzled by the question.

"Sure, but this is merely an observation of what is happening, and says nothing of the difference of states that has taken place. See, we are used to analyzing the process of death as the progressive loss of functions up to total standstill, which is followed by decomposition. But this is only a 'mechanical' method of observing the phenomenon. Now I am going to ask you: what has that man lost one moment after dying compared to the moment before?" insisted Henry.

"Well, life" answered Jan.

"Sure, but let me ask you: did you see it leaving his body? Could you observe the process and watch this 'thing' dissolving or going somewhere else? Where was it a moment before and where has it gone a moment later? Could it maybe be something that 'intersects' our world from another dimension, and therefore is invisible to us, something that has stopped producing effects on that being?"

"Fine" cut in Jan, "going back to the similarities with the flat world, you mean to say that what we call life might be nothing more than a kind of 'hyper-candle' that passes through us and that, as its intersections with our space get thinner and thinner, we see it fade and eventually disappear?"

"Regardless of what the explanation might be, it is good that you begin to imagine the things around you with this approach."

"This brings another question to my mind" said Jan and continued, "I clearly understand that we are able to perceive only three dimensions, however your last point makes me think that this is not sufficient to affirm that we are three-dimensional beings. We could be multi-dimensional beings who are able to perceive only three dimensions."

"Excellent observation" replied the old man, "you are certainly on the right track. Now, if you would follow me, we will take another step forward."

"I'm all ears."

"If you were to assume that man, although he can only perceive three dimensions, is actually a four-dimensional being, then, as you will see, we would be able to derive different assumptions and explanations for some fundamental concepts."

"Like what?"

"If we cross through these dimensions in increasing order, we can certainly affirm that a dot is nothing more than the section of a line. A line is nothing more than the section of a flat plane. A flat plane is the section of a three-dimensional space. Then, we can certainly assume that a solid, as we see it, is nothing more than the section of a four-dimensional body."

"Yes, I understand your reasoning." said Jan.

"Now back to the usual example of the flat man coming to grips with three-dimensional entities. We have seen that he perceives an infinite number of sections of the three-dimensional object. These are nothing but intersections of the solid with his flat plane, occurring over time. But what are, in reality, these intersections if not the infinite number of moments of existence of the three-dimensional body in two-dimensional time?"

"Yes, he sees the sections of a multi-dimensional object over time" agreed Jan.

"So, if the man suddenly became four-dimensional and observed himself, time would disappear for him and he would only see an endless succession of images of his life, a bit as if he was observing living photographs that continue to exist, even though, with his previous three-dimensional limitations, events would seem to be hopelessly lost in the past. At the same time he will also have the ability to see the living photographs of the many ways in which his future can develop and which already exist."

"But then, if the past still exists and the future already exists, this means that our whole life is predetermined and that whatever we do, our future is already preordained", Jan said with a frown.

"Not really. You see, going back to what we saw regarding quantum mechanics, we can say that the past contains not only what has happened but also what might have happened, and likewise, the future contains not only what is going to happen but also what might happen. Therefore, what we do can somehow determine which of the possible futures will come true. But all of them, however, already exist. In fact, if you had the ability, hypothetically, to foresee the future, you would not see what is going to happen, but all that could happen."

"What do you mean?" asked Jan.

"To give you an example, we might compare our condition to that of a driver who is proceeding with an obscured windscreen and who cannot see what lies ahead until a certain event takes place and is physically manifested to him. Obviously he acts on his steering wheel using his common sense and his experience, but he cannot be certain of what lies ahead."

"Therefore he proceeds cautiously but blindly", added Jan.

"That's right. Now, if his windscreen should suddenly became clear, making it possible to see what lies in front of the car, he would not only see the road that he will actually take, but all the possible roads that he could take, just like all the possible futures and the possible conditions allowed in quantum mechanics.

Our driver, therefore, though able to see the future, is still free to act on the steering wheel to direct the car towards the path that he prefers."

"I understand," said Jan, "so with the ability to choose, the concept of free will is saved, in a way. However, all this leads me to think that we have always existed and will continue to exist forever, and that what appears to us to be our life is nothing more than the interval between the moment our senses begin to perceive the surrounding environment until the moment that environment disappears from our 'sight'. And this does not mean that we have ceased to exist, but only that we are no longer able to perceive ourselves."

"Or maybe that we have been released from our limitations and we have begun to perceive ourselves in a different and more comprehensive way, as we really are." concluded the old man, who could not conceal his satisfaction for Jan's progress.

The sun was now setting and the two prepared for their usual frugal dinner.

SIXTH DAY

THE EXPANSION OF CONSCIOUSNESS

Jan woke up late. He had spent most of the night lying on the rug, thinking. All those considerations had been swirling in his mind and, if on the one hand the loss of all his certainties had triggered feelings of confusion, on the other he felt pervaded by a growing calm. Perhaps, he thought, the clear-headed reasoning that demonstrated the absence of time, not only represented a victory over death, but also endowed every moment of his life with enormous value. Everything, every gesture, every experience would never be lost, swallowed up by time, but would remain with him forever.

The old man had already prepared breakfast and was waiting for him to share it.

"How are you?" asked the old man as he approached the table.

"I am confused. It is not easy to wipe out all the concepts and certainties that we have had since childhood."

"Come, let's go for a walk in the forest, you'll see your confusion will diminish." suggested Henry.

They walked towards the forest. It was a beautiful day and the bright sun enhanced the beautiful colors of nature that were typical of that time of the year.

"I understand your confusion. You need a lot of patience and above all concentration. Do you remember when in the beginning I told you that the most important thing was to learn how to go out of the context? That meant getting away from all that we instinctively believe in order to form a new idea of life, of the world and of what we really are." said the old man.

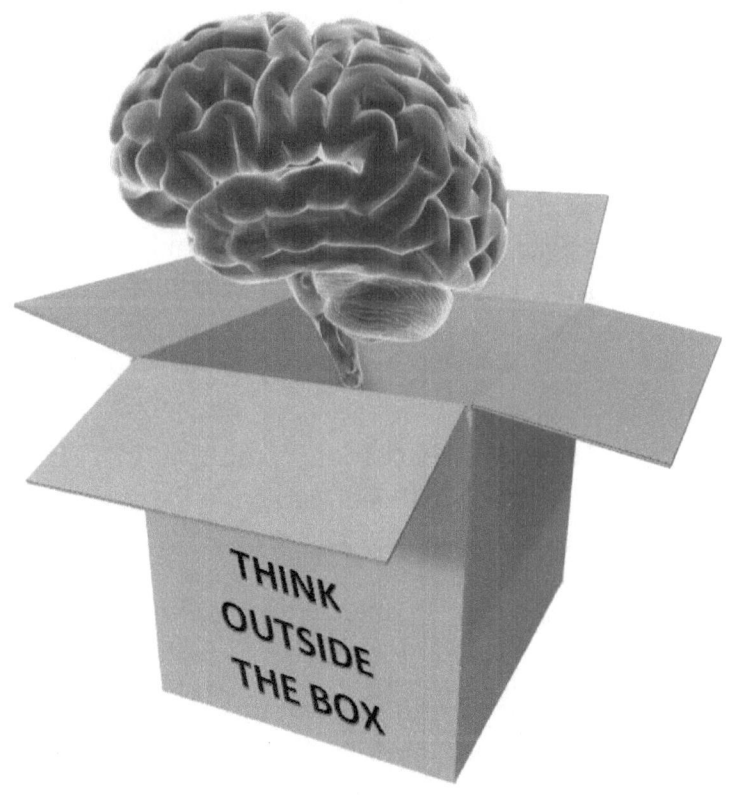

"Yes, and that is particularly difficult" said Jan and he continued, "In fact, last night I was thinking that we have not considered the point of departure from where the development of a higher state of consciousness should start."

"What do you mean?" asked the old man, curiously.

"Well, for example, we have imagined the flat man watching with amazement the appearance in his world of lines and dots that are inexplicable to him. But what we have not taken into account is that any thinking being who belongs to any number of dimensions elaborates his idea of the world surrounding him starting from the observation of his own world. So the appearance of these lines or dots would be perfectly normal for the flat man, although inexplicable, because they would be part of his natural world. Therefore, he would not be so surprised, he simply would not know how to explain a natural phenomenon that has always been part of his world, just like many other phenomena that are part of ours and that we consider natural, although we cannot fully explain their origin. This would most certainly hold back the development of a state of higher consciousness because these phenomena would not be considered anomalies." concluded Jan.

"This is an excellent observation," the old man said with satisfaction. "And it's why, as I said, much concentration is needed."

"Yes. But it is not at all simple," said Jan.

"It is just a matter of wanting it with enough determination," said the old man, who continued, "Would you like to hear a curious anecdote that comes from oriental culture?"

"I would love to." said Jan.

"Well, the story goes that at the court of an important ruler, some courtiers asked him why he continued to send food and clothing as a gift to an old ascetic who lived on a mountain and who never even took the trouble to thank him. The king replied that he did not feel offended by the lack of acknowledgment as the ascetic was so intensely concentrated that he did not notice anything but the object of his thoughts. The courtiers expressed their skepticism and said that it was not possible for anybody to be so focused on something to not notice what is happening around. So the king decided to teach them a lesson. He filled some vases with water up to the brim and gave one to each courtier, ordering them to take a walk around the city and to come back with the jar full of water. Each of them was to be followed by a soldier with a drawn sword whose job was to cut off the head of the poor courtier at the fall of the first drop of water.

The terrified courtiers carried out the order and all of them were able to make the journey without spilling a single drop of water. When they returned, the king congratulated them and asked them to describe the people they had encountered along the way and the things they had seen. No one knew what to say, and they all affirmed that they had been so careful to not spill a single drop of water that they were not able to notice anything else. Then you see, said the king, when you are sufficiently motivated you can be so concentrated on something as to not realize what is happening around you"

"A really good story" said Jan.

"Good, now stay focused and follow what I say." said the old man.

"I will try, but promise not to cut off my head if I am not able to" replied Jan, smiling.

"So, if the flat man should meet another flat man, with a higher level of consciousness, who explained to him that what he considers to be time is nothing but space in a higher dimension, and that his perceptions are misleading and incomplete, would this be sufficient for him to make some progress? In other words, is to become aware of being in a prison also the first step in evading from it?" asked Henry.

"Well, I don't know," said Jan, "but even if the awareness of being imprisoned is not sufficient to escape, it is for sure a necessary condition."

"Now, certainly the first thing the flat being would do, once he comes to realize he is in a sort of a sensory prison, is to try and escape using his senses."

"And is it not the right way?" asked Jan.

"The point is that he would not get very far with that. It is very important to learn how to use all the senses at the same time but only because this helps to develop consciousness. Also in this case in fact, the well-known principle that the whole is greater than the sum of its parts applies. The ability to simultaneously use all the senses will enable consciousness to make a real breakthrough."

"So there is no chance of getting 'out of the box' just by expanding our senses?"

"It's not enough. There is a famous work by Escher that describes the situation from a graphic point of view. It represents a dragon who becomes aware he is being drawn on a sheet of paper, and hence is a two-dimensional being. He desperately attempts to get out and into the third dimension by putting his head into a hole in the sheet and biting his tail, which he had previously thrust through another hole in the same sheet."

M.C. Escher: Dragon

"Very original" remarked Jan.

"Yes, but his attempt is doomed to failure. In fact, those openings are only drawn on the sheet and therefore are just an illusion. He invariably remains a flat being confined to his two-dimensional world."

"I understand. So it is imperative to use conscience and reason to escape from the prison, and not just the senses." said Jan.

"Yes. If the two-dimensional being begins to revise his concepts from scratch and to consider the universe as something that is simultaneously in his past and in his future, he will acquire a higher level of consciousness that will open his eyes to a world of such marvelous complexity and beauty that he will no longer wish nor will he be able to go back to the limited way of thinking he had in the beginning."

"I understand what you mean. It's what is happening to me." said Jan.

"If you succeed in expanding your consciousness up to this point you will be able to perceive a new world. A world devoid of time, where things exist independently from their manifestation in a section of space, like, for example, in our three-dimensional world. In the new world consequences exist simultaneously along with causes, events that appear to be distant from one other in time can actually be next to each other. There will be no conflict between existence and non-existence, life and death, but it will be evident how one contains the other."

"Yes, and it's wonderful" said Jan.

"In fact, everything will appear to you in a new light and with a beauty which has no equals, just as the laws of physics become more and more simple and let us say 'elegant', as they are referred to and expressed in an environment with an increasing number of dimensions. You will feel as though, until now, you had been watching the world through a tiny crack, like an ant who, immersed in the chaotic bustle of its anthill, cannot grasp the high level of its functional organization unless, by some miracle, it is able to observe it from above." concluded the old man with an inspired smile.

"Then, speaking of science, you mean it's likely that one day, all the thousands of equations that describe the operations of the surrounding world could be synthesized and expressed by a single equation which refers to a multidimensional world?" asked Jan.

"Why not?" said the old man and continued, "but there is more. See, one of the most beautiful thoughts that we draw from this is that science seems to point towards concepts that until recently were reserved to contemplative spirits, and referred to philosophy or religion. Thus, they had no significant impact in practical terms and just remained at an argumentative level. For instance, over the centuries we have developed a science that had at its foundation the idea that both the infinitely large and the infinitely small are made of matter."

"Yes, from the planets to the atom" agreed Jan.

"We research every field, from the macroscopic to the microscopic, from galaxies to the molecular, atomic, nuclear and sub-nuclear levels, and as science gradually makes progress, we discover that what we thought to be different entities and phenomena, - gravity and electromagnetism, radioactivity and nuclear forces, quarks, protons, neutrons, time and space - are just different waves of the same energy field. The more we dig into the infinitely small to search for the fundamental 'brick' of matter and life, the more the universe slips through our fingers, until we reach something that becomes more and more abstract and finally immaterial."

"Yes, it's absolutely extraordinary" said Jan.

"Everything seems to have the same nature and is attributable to a single unifying cause. We have understood that matter in itself is nothing but a form of energy, at all levels, and therefore the distinction energy-matter does not make any sense. Quantum mechanics, for example, conceives the universe as made of waveforms and not matter. Yet, as we have seen, its laws are very concrete. They explain our physical world better than in the past."

"Energy, then" said Jan.

"That's right. And that very energy is the same that forms you and me, just as it does flowers, rocks, trees, and so on. We, and everything around us, are made of the same energy, even if this is combined in the many different ways with which it intersects our environment. So it is clear that at the basis of the apparent diversity of life there is a single cause and a single manifestation of energy."

""This makes me think of the creative energy given by God to the universe." said Jan.

"Certainly, science is coming to the same conclusions as those of mystics and believers, or philosophers, for non-believers. And these concepts are not abstract speculations but have actually been experimentally demonstrated."

"Incredible," said Jan, "science and faith begin to converge after thousands of years of separation."

"Yes, but be careful. We cannot say that science is discovering the existence of God, but only that we are now beginning to look outside, glimpsing through the bars of a sensory prison of which we are now aware. What we can see is that science is revealing that the entire universe, including ourselves of course, is composed of various combinations of the same form of energy."

"So you suggest not to abide by the interpretation offered by religion." said Jan.

"Exactly, we are talking about science, and if some of its conclusions happen to converge with religious interpretation, it's certainly something worth reflecting on, but it is also to be kept separate from the scientific debate."

"Why?" asked Jan, amazed by the seemingly useless strength of that statement.

"You see, I believe it is of the utmost importance that these concepts, and more generally, the development of a higher consciousness, become part of the scholastic curriculum, just as the Pythagorean theorem or literary masterpieces are. So far, this has not been the case, except in rare occasions, thanks to the secularist approach of scientific teaching and in order to guarantee neutrality with respect to different religious beliefs. But now that it is science speaking out, now that the idea of achieving a higher consciousness is not based on religious belief but on rigorous scientific experiments there is no longer any reason to exclude the concept of achieving a higher consciousness from school teaching. As you may have noticed during our discussions, science is leading us to conceive the world in a way that is very different from what our senses tell us, and this requires a great effort of understanding, which is sometimes incomplete. And yet our, teaching programs are not planned to open-up the minds of young people, to help them systematically reason and think in a different way. It is not difficult to imagine that, given the width of the conceptual leap, it will take several generations before such a complex reality will be acquired. What's more, I am absolutely convinced that this would give future generations a much broader view of who we are and what our place is in the universe. Most likely, this would also make us less inclined to wage war and show hostility towards each other, and would render all differences based on race, skin color or any other factor that has divided humanity for centuries totally ridiculous."

"Yes, you are right. I too have made a similar reflection and completely agree with what you are saying." concluded Jan.

They walked in silence while Jan reflected on what they had just discussed. The old man's words had a powerful effect on him and his compelling arguments were totally convincing. Yet, he couldn't help but think, just as the old man had said, that all they had talked about remained only an abstract topic, which could not produce any real change if not awareness.

Then, suddenly, something happened in his mind and he was pervaded by a strange, and at the same time beautiful, feeling. During their walk he had been struck by the vivid colors of the forest as bright sunshine broke through the foliage of the trees, he had relished in the intense scent emanated by wild flowers and, in a moment of silence, he had paused to listen to the song of birds flocked in some of the trees. Now, suddenly, all the single emotions that he had felt pervaded him altogether, each one relinquishing its own specific effect, but creating something that was infinitely greater than the sum of their parts: a sort of 'emotional concert', which completely pervaded him. For the first time ever, the whole surrounding environment had entered into him and he perceived it, infused into a single entity, as a unique cognitive experience. It was an inebriating experience. It was no longer sight, hearing or smell that informed his perceptions, but something else, something undefined and yet absolutely real.

He remained in silence, his eyes wide open and a hint of a smile, as he felt an intense joy that until that moment had been unknown to him.

They headed back to the hut, walking slowly and without speaking, each one enjoying the beauty offered to them by nature.

After having consumed a frugal dinner, Jan felt exhausted, his eyelids heavy. He took leave of the old man and lay down on his rug which now seemed much more welcoming. As he stared at the ceiling for a couple of moments he felt all the limbs of his body relaxing, one after the other, and only then realized how much tension they had accumulated. For the first time in ages, he slept soundly.

SEVENTH DAY

THE MISSION

Jan woke up with a thousand questions and an urge for answers.

Turning to the old man he asked, "So reality has always been right before our eyes but we have never correctly perceived it.

"That's exactly so. This is also due to the fact that knowledge was for a long time the prerogative of few, and no real care was taken to spread it." said Henry.

"True, but I don't understand why no one, in the past, realized the importance of disseminating knowledge, as you are doing with me right now."

"You see, one of the greatest misconceptions of the past was to think that most men are not interested in acquiring knowledge that was consequently reserved for few. A major argument supporting this view was that he who pursues knowledge, after all detracts nothing from others; he does not deprive anyone from anything, but limits himself to grasp what would otherwise have been thrown away. In short, according to this belief, the accumulation of knowledge on the part of few would be the result of the refusal to acquire it on the part of many."

"It seems only a baseless justification to me." said Jan.

"Much more than that. I consider it a real crime. He who achieves a breakthrough in the field of knowledge should feel the precise duty to share it with others. Reality belongs to everyone just as science and the acquisition of a higher knowledge are essential to everyone's lives. To give you an example: if after a shipwreck you reached a desert island together with some companions who shared your misfortune and who managed to survive like yourself, and one day you found a spring of water, you wouldn't even think of not sharing your discovery and keeping the water to yourself, would you?"

"Certainly not. I would share the information immediately because water is fundamental for everybody's life" Jan answered.

"And culture and knowledge are just as important. Many wars and violence would have been spared, in my opinion, if we had bothered to spread knowledge as much as possible." said the old man.

"And this also applies to science" added Jan.

"Sure, but as we can learn from history, awareness and expanded consciousness often reach the understanding of truth before science does. Human history is full of such examples, from Greek philosophy which spoke of the atom before science had the slightest idea of what it was, to Einstein who grasped certain fundamental principles of our universe long before they were mathematically and experimentally proven."

"So you think that sooner or later science will experimentally prove what the expanded consciousness of some has already understood?" asked Jan.

"I am going to answer you with a story that could have taken place in our dear flat universe, if it actually existed." said the professor.

As he listened, Jan noticed that Henry's wrinkled face, marked by age, had gradually assumed a more relaxed and serene expression in the last few days. Almost as if the urgency to communicate his truths had eventually been appeased.

The old man went on: "We have seen how the world of the flat being is actually 'creased' and twisted into a direction that is precluded to him, a third direction, which is perpendicular to the other two and which he perceives as time. We know that he cannot be aware of this and therefore he has developed physical laws that work perfectly for him. In fact, he does not notice the inaccuracy of his laws, because the ripples of his world are quite insignificant and the speed in question is very far from that of light."

"Yes, it would be difficult for him to think that many basic concepts in his physics are based only on illusions." said Jan.

"But one day, a young visionary scientist in the flat world, called Flat-stein, sensed that all that 'flatness' was perhaps unreal and that the flat plane might have some 'curvatures' in the direction of what appeared to be time. There followed a great debate among Flat-scientists, as the deviations from the laws of classical flat-mechanics became relevant only at high speeds or in presence of huge ripples. But some scientists had found a way to measure time with great accuracy and, just as it happens in our world, they were able to experimentally demonstrate that Flat-stein's theories were correct.

These experiments were decisive and, if on the one hand they put the skeptics to silence, on the other they sparked the imagination of other flat-scientists who began to produce new theories and to conduct repeated experiments, proceeding on the new path.

It occurred then to some that if their world was actually curved into another dimension in some points, perhaps they would need to investigate those very points. Some even speculated that from those points it would be possible, depending on the curvature, to access the future or the past."

"Right, after all, in the example of the spring, the flat being who casually finds himself on top of a ripple could see something that his colleague who stands on the level of the flat plane would be able to see only later" said Jan. Then he added: "However, I cannot help but wonder how can the flat man identify these ripple points, since he cannot see the direction in which his space is bent into?"

"Of course it would not be easy, because the eventual points could be extremely tiny if measured according to a normal scale in his dimension; a bit as if, in our case, we were talking about ripples as small as a fraction of a millimeter. But we do not know whether in some point these areas can be larger.

As to how our friend may be able to find them, that could depend on anomalies." replied the old man.

"What do you mean?" asked Jan.

"Well, do you remember the example of the strange forces that he experiences when, without knowing it,   he walks up and down a hill?"

"Sure."

"That is an example that I purposely exaggerated for the sake of better clarity, because, in his reality, these forces were so small, they could be neglected by his classical mechanics and would only became relevant in the case of larger deformations or higher speed. So until Flat-stein postulated their existence and until extremely accurate flat-tools were invented, it was impossible to verify experimentally."

"And then… What could the flat man do?" asked Jan.

"For instance, he could investigate anomalous zones in his space where he could even repeat some of the most fruitful experiments, a bit as if we were to repeat the experiment of the atomic clocks in areas of our world where we found anomalies in the force of gravity or in magnetism. Maybe, who knows, our Flat-scientist could be so lucky as to find a particularly intense and large 'curvature' where less energy is needed to cause a shrinking of space-time, which would enable him to produce the phenomenon using the technical instruments at his disposal."

"That is true." commented Jan.

"Of course we too, perhaps, should further investigate these strange anomalies of gravity and time. For example, in the Hudson Bay area in Canada, a significant decrease of gravity has been verified in some areas. It has been speculated that this may be due to the simultaneous presence of thick ice and of major masses below the earth's mantle."

"Interesting" Jan said and continued: "However, I don't think the search for abnormal areas would be very easy since over 70% of the earth's surface is covered by oceans and we cannot exclude that the most favorable areas could be in open sea."

"True. I am pleased with the progress you are doing in achieving a higher state of consciousness, so it's time for you to come with me to a special place," said the old man.

Jan followed him lazily, absorbed in his thoughts. They walked along the hillside until they reached the entrance of what appeared to be a cave.

"Come, let's go inside together," said the old man.

At that invitation, Jan stopped chasing his thoughts and turned his attention to the surrounding environment. The two crossed the entrance of the cave. After making their way through a few meters of a dark and twisting tunnel, they turned a corner and Jan saw a bow of shining light.

"Don't be afraid, let's go inside" said the old man.

As soon as Jan entered into that vivid glare, he could hardly believe his eyes. He was surrounded by an infinite, 360 degree myriad of objects and people, as if in a kaleidoscope.

It felt almost as though he was navigating in a void, as the images were located at various distances, and were both above and below him. Jan walked slowly within that strange environment and, as he advanced, the world around him seemed to increasingly change shape. At one point, a figure materialized; it was small at first, but it became larger and larger and it stopped in front of him. It was another man and he had his back to him. As the figure grew larger, Jan noticed that in front of him the figure of another man with his back to him was also growing, then it became two figures, then three, and so on until an almost endless array of people formed, one behind the other.

Jan was more and more amazed and wanted to interact with the person in front of him, to ask him what kind of environment they found themselves in. He reached his arm out towards the man standing in front, but as soon as his hand touched the man's shoulder, he felt a touch on his own shoulder. Instinctively, he turned his head to see an arm and a hand resting on his shoulder.

Even more surprised, Jan turned his head further to see whom that arm belonged to, but he could not because the arm simply emerged from the light and he could not perceive the body it belonged to. So he withdrew his arm from the man standing in front of him, and simultaneously, the hand resting on his own shoulder was pulled back.

Incredulous, Jan repeated the gesture several times but the result was always the same: when he stretched his arm towards the shoulder of the man in front of him, another arm materialized from behind and touched his own shoulder. It took Jan a while to understand that it was his own arm and he said: "How is this possible? You have brought me into a hypersphere?"

He could hear the old man's voice who laughingly replied: "Good, that's enough for today. I am sure you will reflect on this for a long time and that it will be very enlightening for you."
The old man led him outside and they walked back to the hut.

Jan dreamt all night. An uneasy dream. He had touched an infinite number of replicas of himself, all of them present at the same time. Himself, another him, then the old man laughing. The images of endless Jans swirled in his mind, piling into an uninterrupted series of images. Until all of a sudden, he understood. He woke up with a start and lay, motionless, with eyes wide open for a long time. Then he got up and called out to the old man, but no one answered. The old man was gone, he had freed his goats, who were now grazing on the hill.

So, the crazy old man was actually himself. He had travelled back into his past in order to meet himself as a young man. In Jan's mind everything was now in place: the logo of his signature, the briefcase and his knowledge of the secret code to open it. Evidently, in a not too distant future, time travel had become possible. Well, thought Jan, as time does not exist, it had been a journey into another dimension, which allowed the traveler to reach some region in space that was otherwise precluded. But why had the old man done all that? And why was he gone now?

As he pondered on all of this, Jan rummaged through the things that Henry had left inside the hut. He found a rather old and crumpled University ID card made out to a Professor Jan Korpinsky; he was surprised to see a photograph on it of his own aged face. At that stage, it was easy for him to recognize himself, but because he had not recognized the old man, that evidently meant that the journey back in time had occurred several years after that photo had been taken.

Lost in reflection, Jan sat outside the hut for a long time. The infinite consequences the story presented complicated his thoughts. However, if thanks to this incredible encounter his knowledge and his level of consciousness had increased significantly, this would also have benefited the old Jan, who had now gone back out there, somewhere, to a space that to him seemed to be in the future.

Suddenly he remembered the words he had heard from the people at the restaurant the day he arrived. The old man was called Henry because many years before another old man with that name had mysteriously appeared on that very hill and lived in the same hut, and he too, after some time, had mysteriously disappeared. It was not the first time, then, that the crazy old man had turned back on his way. But why? What was his purpose?

The answer came to him, but not through reasoning. He could feel it coming from within, from his consciousness, although he could not explain how. Henry was pouring his knowledge into his young self, as if he were just a container, thus increasing his own knowledge every time. Then, he would go back again and repeat the same process.

But then, Jan wondered, why did he not have any memory in his life of anything that had already happened? Which Jan, among the many, was now asking this question? There could only be one answer. Each time Henry picked a parallel Jan from one of the possible states of the past and allowed him to live and to evolve by pouring a higher level of knowledge into him. After all, as Feynman affirmed, each system does not present just a single story, but every possible story amongst those allowed.

Then, Henry's journey was nothing more than a victory over death and over time, which imprisons us and places a limit, the limit of life on our ability to expand our consciousness and knowledge.
Jan picked up his things and walked towards the village. His life had changed forever and nothing would ever be the same.

The state of his consciousness allowed him to feel that his life and, ultimately, the history of the world, would continue to exist forever, somewhere, as they always had, and even though he could not see those spaces and interact with them, the simple awareness that they existed, took whatever power away from death and gave him an inner peace that would never leave him again.

He turned to look at the hill one last time as the sun was setting, and his lips murmured: "Thank you."

BIBLIOGRAPHY

Stephen Hawking: The Nature of Space and Time - Princeton

Stephen Hawking: A Brief History of Time - Bantam

Stephen Hawking: Dal Big Bang ai buchi neri - Rizzoli

C.H.Hinton: The Fourth Dimension - Celephais Press

D.Hofstadter: Godel, Escher, Bach - Adelphi

Michio Kaku: Hyperspace - Oxford University Press

Ouspensky: Tertium Organum - Astrolabio

Ouspensky: The Fourth Way - Astrolabio

Abbott: Flatland - Eldritch Press

F.A.Wolf: Anelli temporali e torsioni spaziali - Macroedizioni

P.Davies: I misteri del tempo - Mondadori

ABOUT THE AUTHOR

ICT specialist and author of many books and publications of technical/educational nature.